THE ANGRY CHEF'S
GUIDE TO SPOTTING
BULLSH*T
IN THE WORLD OF FOOD

THE ANGRY CHEF'S GUIDE TO SPOTTING BULLSH*T IN THE WORLD OF FOOD

BAD SCIENCE and the TRUTH about HEALTHY EATING

WITHDRAWN

ANTHONY WARNER

THE EXPERIMENT

NEW YORK

Library of Congress Cataloging-in-Publication Data

Names: Warner, Anthony (Chef), author.
Title: The angry chef's guide to spotting bullshit in the world of food : bad science and the truth about healthy eating / Anthony Warner.
Description: New York : The Experiment, 2018. | Originally published in Great Britain as The Angry Chef by Oneworld Publications in 2017. | Includes bibliographical references.
Identifiers: LCCN 2017043524 (print) | LCCN 2017053572 (ebook) | ISBN 9781615194612 (ebook) | ISBN 9781615194605 (paperback)
Subjects: LCSH: Nutrition—Popular works. | Diet—Popular works. | Food—Popular works.
Classification: LCC RA784 (ebook) | LCC RA784 .W3635 2018 (print) | DDC 613.2—dc23
LC record available at https://lccn.loc.gov/2017043524

ISBN 978-1-61519-460-5
Ebook ISBN 978-1-61519-461-2

Cover and text design by Sarah Smith
Author photograph by Image North

Manufactured in the United States of America

First printing April 2018
10 9 8 7 6 5 4 3 2 1

To endure uncertainty is difficult,
but so are most of the other virtues.
—BERTRAND RUSSELL

Contents

Prologue ix

Part I: Gateway Pseudoscience 1

   1. Confused by Correlation 2

   2. Detox Diets 15

   3. The Alkaline Diet 31

   4. Regression to the Mean 43

   5. The Remembering Self 57

Part II: When Science Goes Wrong 67

   6. The Genius of Science Columbo .68

   7. Coconut Oil 89

   8. The Paleo Diet 106

   9. Antioxidants 124

   10. Sugar 135

Part III: The Influence of Pseudoscience 153

   11. A History of Quacks 154

   12. The Power of Ancient Wisdom 165

   13. Processed Foods 175

   14. Clean Eating 191

   15. Eating Disorders 203

Part IV: The Dark Heart of Pseudoscience 219

   16. Relative Risk 220

   17. The GAPS Diet 234

   18. Cancer 251

Part V: The Fight Back 273

    19. The Evolution of Myths 274

    20. Science and Truth 285

    21. Fighting Pseudoscience. 295

Epilogue 307

Appendix 1. 311

Appendix 2 312

Acknowledgments. 316

Notes 319

Prologue

*H*ello reader. Thank you for choosing this book. Clearly you are a very sensible individual with an interest in food, looking to learn something new. That's good, because I am a chef with a passion for cooking, a background in biological science, and a fascination with the way our diet affects our health.

Perhaps you are hoping to read about a single hidden secret to healthy eating, or the key to sustained weight loss. Maybe you are looking for a list of ten essential, health-transforming superfoods that you need to include in your diet. I do wish I could provide you with these things, a few simple rules and easy solutions—wouldn't that be nice?—but unfortunately life is just not that straightforward. If it was and I knew all the answers, I would most likely be driving up to my brand-new yacht in a solid gold Ferrari.

In this book, you will not find a list of rules to follow for a happy, healthy life. I will not attempt to break down common foodstuffs into lists of things that either cause or cure cancer. In fact, if this book achieves its goal, it is likely to leave you knowing less about the science of food than you do right now. Or at least less than you think you know.

The Angry Chef was first revealed to the public in 2016, but he can be traced back to a couple of years before. I am a UK-based development chef with an interest in nutrition and was attending a food and health industry event at a large London conference center. There was a panel discussion on the subject of "What is healthy eating?" and I noticed that a then little-known health blogger and Instagram star was due to appear. I was vaguely aware of her and the "clean-eating" trend she represented and was interested to hear what she had to say. Even an out-of-touch technophobe like me couldn't ignore the fact that, in this modern information age, the rise of online stars has the potential to profoundly affect the behaviors and beliefs of the millennial generation. The fact that a number of these new stars were focused on eating healthily seemed encouraging.

It did not take long for me to start worrying about the sort of unregulated advice that some of these new stars might be spreading. Although the blogger in question was likeable, intelligent, even informed in some areas, some of the things she was saying were a little strange. At one point she claimed that anything cooked at home is bound to be healthier than something made in a factory, which I expected to raise the hackles of many in attendance, especially as this was a food industry event. As I looked around, the audience seemed to be nodding sagely, in complete agreement with her argument, and I felt for a moment as if I was trapped in some late 1950s sci-fi movie, a dystopian future where I am the only one in the crowd who can see the false prophet.

I left the event slightly confused, but not yet particularly Angry. I was curious about some of the strange beliefs of the clean-eating movement and started to do some research. The more I read, the more incredulous I became at the mangled misunderstanding of science and the absolute nonsense that underlies some of these trends.

Ever since, I have been down the rabbit hole, transported into a world of strange pseudoscience, arbitrary rejection of modernity, and dangerous dumbfuckery that has come to dominate the discussion on food and health. Clean eating started as a fringe movement but has grown into a huge and unrepentant tide of nutri-nonsense. The health and wellness lobby is taking over, relegating the opinions of nutritional scientists, dietitians, and public health officials to the sidelines. Their books dominate the bestseller lists, their websites receive millions of hits, and their Instagram accounts deliver endless pictures of kale smoothies and quinoa bowls to armies of adoring followers. The more I look, the more my eyes have been opened to this bizarre and sometimes dangerous world: a world where lies are told about food every day.

In the two years since that event, my fascination and revulsion at clean eating have moved on considerably. In many ways that particular blogger was the more respectable tip of a dangerous, unregulated, and ever-growing iceberg. As the trend soars in popularity and new stars fight for space in an increasingly crowded market, the advice being doled out by poorly qualified and unaccountable fools has had a tendency to become more and more extreme.

The one thing that unites these voices is that they all spread their message with great certainty. When they say that lemon water is alkalizing to the body, the reason that people buy this lunacy is that they are so very certain that this is the case. Look into their eyes and you can see that they truly believe. Throughout this book we investigate some of their claims and try to understand what underlies these false beliefs and how they have managed to become so popular.

As the wellness movement has grown, it has become increasingly powerful. Sometimes it seems that there are precious few moderating voices in a world gone mad. To make matters worse, the madness no longer stops at health bloggers. As we shall see, bandwagon-jumping celebrities, medical doctors, and even an increasing number of specialized academics are susceptible, blinded by the light of the wellness stars and desperate to serve our insatiable demand for certainty.

Unlike the purveyors of pseudoscience that I rail against, I offer you no certainty, no easy answers, no simple stories. In food science, as in all science, progress often depends not on the conviction of experts, but on our ability to accept what we don't know. This book is an investigation of bad science in the world of food, and part of that is understanding that there are gaps in our knowledge. Unfortunately, the nature of our minds can make this a very difficult thing to do.

The Bertrand Russell quote at the start of the book is one of my favorites and underlines many of its messages. Sometimes in life we have to endure uncertainty and act in the absence of evidence. Although the science around food and health is complex, we still have to eat every day. To have a healthy relationship with our food, we need to learn to accept that we don't know everything. I am not saying that we should just shrug and eat jelly bean sandwiches or deep-fried cake for breakfast without thought of the consequence. There are some clear links between food and health and our dietary choices have an impact on many serious diseases. But we have to look at the evidence, and often the available evidence is not good enough for anyone to get up in the pulpit. Which brings us to the first piece of the puzzle . . .

Part I

GATEWAY

PSEUDOSCIENCE

Chapter One

CONFUSED BY CORRELATION

TIME TO THINK

If there is one thing that I hope this book will encourage people to do, it is to take some time to think. Modern life leaves us bombarded with information, and it can become too easy to live our lives making only quick instinctive judgments. Our instincts can lead us down treacherous paths at times, especially when it comes to decisions about our diet. Taking a few quiet moments to process the barrage of information we are subjected to every day is perhaps the greatest weapon we have when it comes to fighting bullshit in the world of food.

I have been lucky in my life. For many years, my work as a chef involved a great deal of menial and repetitive tasks. Cell phones were banned from most of the professional kitchens I worked in (I know because it was me who banned them) and other distractions were limited, as you are unlikely to be able to get the job done without a lot of focus. Cooking for large numbers of people is generally a slow and consuming process—and that gives you plenty of time to think.

As my career developed and the nature of my work changed—and as information technology became even more integrated into our lives—the slow, contemplative moments in my day dwindled. These days, it is rare that I will have fifty sea bass to fillet, or three cases of baby leeks to trim. To make matters worse, upon entering middle age I have had to reluctantly engage with social media, plunged into a world where vast quantities of information are delivered in a constant stream of needy clickbait headlines. Every day they demand my limited attention like a nest full of starving chicks begging for food. Like the vast majority of people, I am subjected to a shitstorm of emails, messages, newsfeeds, headlines, images, notifications, timelines, newsletters, Skype calls, 24-hour news channels, and advertisements. All are increasingly tailored to specifically meet my every whim and desire, fiercely, noisily, and colorfully battling for my attention. I am subjected to countless thousands of pieces of information every day and am constantly forced to make instinctive judgments on every one of them. Should I ignore, engage, share, or react? Should I be outraged, amused, disgusted, empathetic, joyous, fearful, or angry? I must decide in a few seconds, and then move on; otherwise I might drown in the sea of information.

Although such instant and unrestricted access to the world can be powerful and liberating, it is perhaps the great paradox of our age that as we receive more and more information, we seem to become less and less informed. This is the "paradox of choice," and it blights our modern world. This is never worse than in the world of food, where a huge proliferation of often conflicting information leaves us struggling to know what to believe. Many give up, many make quick instinctive judgments, and almost all of us will get some things wrong. If there is one thing that might help us make better decisions, it is taking a little time every day to stop and think.

That's why I run. I get up stupidly early every morning, early enough to allow me a bit of peace each day before the information blizzard overwhelms me. Before I am properly awake, I crawl out of bed, put on my running shoes, and tread a long, familiar route around the fields and paths near my home. With bleary eyes, wild hair, and a slightly deranged springer spaniel in tow, I slowly plod my aching, middle-aged knees and ankles around field and forest, in rain, wind, snow, hail, ice, or sun. I don't love running because I'm competitive or want to be healthy. Some people say that exercise is boring, but for me the boredom is the enticing thing about it.

THERE'S NO SUCH THING AS THE EASTER LAPWING

For much of the year it is dark when I leave the house. Keeping up my daily runs through the winter months is tough, but I enjoy the solitude, the night sky, and the quiet of the hours before dawn. Even so, as spring approaches, seeing the rising sun in the mornings does bring joy. The slow pace of rural England's changing landscape through the seasons is a welcome contrast to the rapid flurry that awaits me when I get home and turn on the laptop.

With the spring comes new life. During March and April, in one particular part of my run, I will start to see the usually elusive hares across the open fields, growing in confidence as they do their crazy March hare thing. In scenes reminiscent of some local towns on a Saturday night, a single female can be seen surrounded by groups of increasingly frantic males, approaching her one at a time, only to be shooed away by deft boxing skills. Consumed and blinded by desire, they will sometimes allow me to get within twenty feet of this display, even when I am accompanied by my lolloping canine companion. Sometimes, very occasionally, I will be lucky enough to see one of the hares sitting next to a pile of colorful eggs set into a little scrape in the ground.

Throughout medieval Europe, legends were born about hares bringing forth these colorful eggs as gifts to celebrate the coming of the spring. These myths persisted, and it is easy to see why. Hares start to appear in the open fields in springtime. They frolic and mate openly. At the same time, in exactly the same fields, crude scrapes in the ground appear, seemingly made by scratching paws, and these scrapes are filled with eggs. Surely the hares are responsible for the eggs.

This story is so persuasive that it has become written into our culture. Although the frolicking hares evolved into a slightly menacing man-size bunny rabbit and the eggs became cheap foil-wrapped chocolate nestled into branded mugs, the myth endures every Easter. But it is based on a misunderstanding. As we hopefully all know, hares do not lay eggs.

The eggs that I see on my morning run are instead laid by the slightly more elusive lapwings. Although predominantly wetland birds, lapwings are springtime visitors to the fields along my running route and throughout much of England. When it comes to laying their eggs they favor the same open pasture that the hares inhabit. Lapwings are, however, considerably more flighty than their leporine neighbors, and are usually long gone before dogs and middle-aged joggers get close.

It is easy to see how people could have been fooled (or at least how easy it would have been to spin a pleasing story to a naive child). To use a scientific concept, the hares and the eggs are closely correlated. We see a big obvious hare next to a bright, shiny pile of eggs and in our minds there is nothing else around that could have produced the eggs. The eggs happen to be of a size that might have been laid by a hare. The hare is sitting next to them, shouting, "Look at me, I am a massive fucking hare." The scrape in the ground looks like it could have been made by hare-like paws. We are quick to discount any other unseen possibilities and our minds are instinctively drawn to create a story, to fill in the blanks, and to jump to conclusions. Once we make that jump, the stories we create can enter our belief system. When we have seen something with our own eyes, we will often believe it from the bottom of our soul.

It is human nature to see correlation and imply causation. The reason that correlation can occur between two things without there necessarily being a causal relationship is explained by something known as a "confounding factor"—the real, unseen cause of the correlation. In this case, the confounding factor is the coming of spring, which causes both the hares and the lapwings' eggs to appear in the fields at the same time. Identifying confounding factors is the key to explaining how closely correlated phenomena might not always be causally linked, but this is often a tricky thing for our brains to do. We are wired to explain the world in terms of the information readily available to us.

Understanding that correlation does not always imply causation is to my mind the most important thing that science can teach us. Throughout this book, I will present dozens of examples of mistaken beliefs and pseudoscience, most of which exist and proliferate due to this misunderstanding. Or rather they exist because of our brain's instinctive desire to create a story out of what it sees.

Although we may laugh at the follies of the medieval folk who believed this to be the case, we are all inclined to mistake correlation for causation and miss potential confounding factors when the story fits our view of the world. The most important thing I can urge you to do is take a moment out of your day to think about the stories you hear. Stop for just a moment and wonder whether or not the new miracle diet or new superfood you have been told about might actually be a mischievous hare sitting next to a pile of colorful eggs.

THE STRANGE CULT OF GLUTEN-FREE

To understand how easy it is to confuse correlation with causation when it comes to diet, we shall need an example. Let's take the gluten-free diet. Probably a good idea to get this one out the way early on. Celiac disease is a horrible thing, a nasty autoimmune condition where the presence of even tiny amounts of gluten in food can cause colossal harm to the health of sufferers. In celiacs, not only does exposure to gluten cause horrendous damage to the intestinal lining, but it may also increase the likelihood of developing certain cancers. To make matters worse, avoiding gluten is seriously fucking annoying and requires huge changes to people's diet and lifestyle. Completely cutting out gluten is socially difficult and expensive and requires care, planning, and professional help to ensure that those afflicted by it do not end up with dietary deficiencies caused by the restrictive nature of the regimen they are forced to follow.

Over the past few years, something quite curious has occurred. A number of nonceliacs seem to have decided that the gluten-free diet is a fun new lifestyle accessory that they should try. A dietary myth has been created around gluten, and huge numbers of people are needlessly cutting it from their diets in the misguided belief that doing so puts them on a path to better health. We shall look at the reasons underlying this later in the book, but suffice it to say gluten is now seen by many to be a great dietary evil. Not just for celiacs, but for all who consume it.

Why should this be? When we look at the clean-eating and wellness movements in more detail in chapter 14, we will see that there are some dark and complex reasons behind this fallacy, but for now let us look at a single, imaginary case study.

Introducing Jamie

Jamie is a young man who is concerned about his health and slightly worried about his weight. He has heard a bit about gluten and has a vague understanding that gluten-free products are somehow healthier. Like most of us, his understanding of the science behind dietary choices is fairly limited. Oliver, his gym instructor, tells him he should try a gluten-free diet, so Jamie decides to take the plunge. He does a little research and decides to cut out all bread, pasta, and pizza, chuck a lot of food out of his cupboards, start reading the labels of everything he buys, and stock up on gluten-free products from the local supermarket.

A couple of weeks later, Oliver asks how the whole gluten-free thing is going. It has been expensive and a bit of a pain, but Jamie is quite pleased. He has lost a couple of pounds, he feels less bloated, he maybe has a bit less wind, he hasn't had a cold for a while, and that patch of eczema on his elbow seems a bit better.

"There you go," says Oliver with the sort of smug, self-satisfied smile only personal trainers have perfected, "that'll be the gluten."

Jamie has given up gluten. He has lost some weight and feels a bit better. He made a specific intervention and has seen an effect. His mind will be strongly inclined to reach the conclusion that the action he has taken has directly caused the result. Jamie now believes that cutting out gluten has improved his health. For him, there is clear evidence that gluten was doing him harm.

But are the improvements Jamie has reported enough to reach this conclusion? In this case, is gluten an egg-laying lapwing, or just a big, stupid, obvious hare sitting next to a colorful nest? Has Jamie missed any confounding factors that might provide a different explanation?

The answer is that we don't know. Although we cannot definitively say that cutting out gluten has not been of benefit, we are also in no position to say that it has. Jamie has done many more things in the past couple of weeks than just cut out gluten. Gluten is just one small protein among the thousands of chemicals contained in wheat flour and the many, many more that constitute a pizza. He has made dramatic changes to his diet, taking out a lot of the staple foods he usually eats. This could well have resulted in him consuming fewer calories, causing the weight loss. He has started reading labels, maybe leading him to take a bit more care about his diet generally. Perhaps he used to eat industrial quantities of bread and pasta every day, so cutting them out has helped with bloating and wind. Maybe his eczema just flares up periodically and has faded naturally in the last few days.

To say that cutting out gluten has caused the improvement is like being confronted by a whole field filled with thousands of different animals and saying that the hare sitting quietly at the back definitely laid the eggs. The reason Jamie has jumped to his conclusion is that a narrative was created in advance, leading him to discount the many other possibilities. If Jamie had never heard of gluten but noticed some health improvements after cutting out pizza, he would not suddenly declare, "Ah, I must be feeling better because I have an intolerance to a small protein in wheat flour that helps give the crust structure," yet as this potential fallacy has been planted in his mind in advance by a well-meaning fitness professional, that is the conclusion he will jump to. It does not matter if that information comes from someone with no qualifications in nutrition. When someone correctly predicts an outcome, we will be inclined to believe their explanation.

What has happened with Jamie cannot be classified as a controlled experiment. It is also not a suitable test to diagnose gluten sensitivity. To see whether Jamie has a problem with gluten, we would need to excise that, and that alone, from his diet. He would need to eat exactly the same amount of gluten-free pizza, gluten-free bread, and so on, to ensure that nothing else was affecting his health. To draw any firm conclusions, we would also want to look at real, measurable markers of health rather than vague self-reported symptoms.

Caused by a simple narrative

Our brain is always very keen to draw conclusions and create simple stories to explain the world—often without sufficient evidence. Throughout human history this has been of great use in helping the survival of the species, but it can cause problems when it comes to making informed decisions. If we have a choice between a dour, white-coated dietitian, who might say something like, "It's complicated, the changes you have seen are largely self-reported and could be due to a large number of factors, including regression effects, general changes to your diet, or perhaps some other undiagnosed intolerances," and the smiling gym instructor proudly declaring, "That'll be the gluten," our brain will be drawn to believe the simple message, even when it is delivered by someone we know to be far less qualified and knowledgeable.

So, why exactly do we find it so easy to make poor choices in these situations? Often, when considering how we make decisions, it is useful to use a two-system model of the mind to explain our frequent departures from logic. In the world of Angry Chef, I sometimes crudely illustrate this in frequent conversations with my strange inner voice.

Hello. Where am I?

Ah, hello. We appear to be writing a book.

Really? Us? Writing a book? How did that happen? You do know you are just a chef?

What's wrong with that? Lots of chefs write books.

Yeah, but they write cookbooks. Are we doing a cookbook? Will it have photos? People like Korean food, you know. And barbecues. We could put in some barbecue tips.

No, this is going to be a book about nutritional pseudoscience. We are going to expose some dietary myths and look at why people are drawn to believe strange things in the world of food.

So, it's going to be like that bit at the beginning of all the cookbooks, where chefs talk about their journey and their food philosophy.

No, it's not going to be like that at all. For a start, we certainly do not have a food philosophy. In fact, **Rule Number 1 in the Angry Chef's Guide to Spotting Bullshit in the World of Food** is: *Never trust anyone who claims to have a food philosophy.*

OK. Well, you know I like taking down health bloggers. Who's up first? Vani Hari? David Wolfe? Kris Carr? Oh, please let it be Carr.

No. We are still in the first chapter, trying to explain some of the reasons why people are inclined to adopt false beliefs.

Oh. There won't be any statistics, will there? You do know everyone finds statistics boring?

No, there won't be any statistics. Not yet, anyway.

OK, good. Still not quite sure why I am here. I have better stuff to be doing, you know. We have a new pack of cookies downstairs, and since Jamie isn't eating gluten, we'll be eating all of them!

Well, Jamie is imaginary and won't be eating any cookies, and you are just a voice in my head, here to explain why he might get stuff wrong.

Oh, sorry. But does it really matter if he gets stuff wrong? If he feels better and has lost a bit of weight, then surely that is the important thing?

Maybe, but think about this for a moment. Cutting out gluten means cutting out half of Jamie's diet. It is potentially a dangerously restrictive approach and Jamie is undertaking it with little under-standing and no professional help. Wheat is a valuable and healthful source of nutrition that forms an important part of many people's diets. Although people often talk about ditching bread because it's "full of carbs," it has the highest protein content of any staple food apart from soy and is a significant contributor of fiber and B vitamins.

Jamie is also cutting out many delicious items, denying himself moments of great joy. For many people, freshly baked bread is the finest food there is, an alchemic mix of a few simple ingredients that can inspire true artisans to create things of great beauty. Equally, well-made pasta is one of the great culinary pleasures, the cornerstone of one of the world's most important cuisines. The same can be said for pastries, pies, pizza, noodles, croissants, shortbread, brioche, and Yorkshire puddings. Although many might consider such denial as trivial, we often underestimate the power of simple pleasures to enrich our lives and improve our well-being.

And although Jamie might have a reasonable diet at the beginning of his gluten-free journey, there is every chance that as the weeks go on, he will adopt different habits. Just because options are gluten-free does not mean they can be eaten with impunity. There are plenty of nutritionally poor gluten-free options out there, and some of the initial benefits Jamie has seen may well fade over time. As gluten-free options are often higher in fat and sugar than their gluten-containing equivalents, Jamie may well be eating a less healthy diet than before.

Perhaps the most dangerous thing of all is the false belief system that has been created. Jamie has accepted the idea that cutting out certain foods is a way to benefit his health. This is likely to become deeply ingrained within his psyche, and in the future, perhaps when the initial success of his gluten-free experiment has faded, he will look for further restrictions, in a vain hope that this might benefit his health.

When exclusion and restriction are unnecessary, they are the exact opposite of what we should be doing to maintain our health through our diet. When we have to accept uncertainty, and when the cause and effect between specific nutrients and health outcomes is largely unclear, the key is to embrace as much variety as possible.

The pull of the gluten-free message is that for celiacs there is a very clearly defined cause-and-effect relationship between a specific nutrient and a specific health outcome. This message strongly appeals to our desire for simple narratives. Many people are drawn to this, and despite the diet being difficult, annoying, and distinctly croissantless, they follow it religiously. These are people desperate to exert control, desperate to make specific interventions, desperate to create certainty in an uncertain world. These are distinctly human desires, and ones that we shall see many times throughout the course of this book.

How to hunt for lapwings

Often we mistake hares for lapwings to satisfy our brain's desire to jump to conclusions. Lapwings are usually more elusive, much harder to spot, and distinctly difficult to pin down. But they can be found. Since medieval times, when hares were thought to deliver gifts to celebrate the season of new life, science has developed many wonderful tools to allow us to discount hares, spot confounding factors, and know when we have found a lapwing. Throughout this book, among the swearing, insanity, and vociferous demolition of pseudoscientific myths, I hope to introduce you to some of the methods that science uses for this and spread a little of my passion for the beauty and knowing uncertainty that they provide.

It is my belief that the scientific method is mankind's greatest work. It gives us the ability to distinguish between hares and lapwings, to discount easily believed narratives, to embrace uncertainty, and to constantly search for the truth. It also allows us to overrule the impulse to believe simple narratives, putting our more considered and reflective self firmly in charge. It was only when we learned to do this that we truly started to progress. Once we had managed to discount the hares and find the lapwings, it took only a few hundred years before we had wiped out smallpox and flown to the moon.

Chapter Two

DETOX DIETS

Now it is time to start looking at some commonly held false beliefs about food. I am aware that for many this is where things might just become a little uncomfortable. Many of the beliefs I am going to discuss are so widely held and so often repeated that they have turned from vague, abstract ideas into perceived reality.

THE INSTINCTIVE BRAIN

I blame our instinctive brains. Our instinctive brains love anything that society and the media have approved, believing thoroughly in the wisdom of the crowd and being hugely susceptible to the influence of bright, shiny celebrities. He is drawn to simple narratives (and caused Jamie to completely give up croissants) and thoroughly dislikes complexity or randomness. He wants to make sense of the world, to believe that someone or something is in control and that everything has a cause. He can lead us to believe some fairly odd things and creates biases so strong that we don't even stop to think.

(As a side note, I hope I am forgiven for referring to the instinctive brain as "he." The only one that I have any real insight into is my own, and he is definitely male. I wouldn't want to upset him by referring to him as "it" all the way through the book. If I am honest, I do automatically think of anything so driven by instinct and so prone to rash judgments as male, but maybe this is my own cognitive bias.)

You are here because I want to introduce people to the idea of the instinctive brain. Our minds can be thought of as being governed by two systems, often operating in conflict. The instinctive brain is the part of us that acts quickly, often with little conscious thought. In Richard Thaler and Cass Sunstein's hugely influential book *Nudge*, they refer to it as the Homer Simpson brain, after the impetuous cartoon character, prone to rash judgments and actions. In general, the instinctive brain is not quite as chaotic as Homer, as this system governs much of our day-to-day life and is responsible for us being able to navigate the world as effortlessly as we do. The instinctive brain is behind lots of the stuff that we do automatically. It tells us when we are hungry, it can tell us not to eat something that tastes bad, and, with a little practice, it can spell, type, drive, ride a bike, and do a little basic arithmetic. It does all these things instinctively, and its ability to perform tasks is often not under our conscious control.

The instinctive brain is also incredibly powerful, capable of making decisions and judgments beyond the reach of even the most sophisticated computers. It can recognize signs of danger before they are apparent and give us the energy and impetus to escape. It can tell if Mrs. Angry Chef is annoyed with us (or, to be more accurate, when she is "not annoyed, just disappointed") from the slightest changes in the tone of her voice, even if she is on the phone and has only just said hello. It can accurately judge in a few milliseconds if someone we have just met likes us, and can even pick up many subtle, near imperceptible clues to give away when someone might be lying. More than anything, the instinctive brain allows us to navigate the world without having to process every bit of information, without having to analyze every interaction, without having to make informed and considered decisions every time we act. It allows us to live much of our lives without conscious thought and so gives the other part of us, our reflective brain, much-needed time to think.

Really. Wow. Get me. How come you are always telling me off for getting us into trouble then?

The reflective brain, on the other hand, is the part of our minds that we think of as being "the real me." It is our conscious self. It governs our most important long-term decisions, thinks about our relationships, considers our dreams for the future, and makes plans as to how we might achieve them. The reflective brain is the one that is reading this book right now, the one that might have an interest in statistics, and the one that carries out any complex cognitive tasks we need to do. Thaler and Sunstein describe this as the Spock brain, after the cold and logical character from *Star Trek*, although this oversimplifies the workings of the reflective brain, which is capable of much more than just cold logical thought.

So, we're a team. An unstoppable pseudoscience-fighting duo. I am your loyal sidekick with special superpowers to drive fast cars, spot danger, and identify disappointed spouses.

Because of the power of the instinctive brain, a lot of false beliefs are firmly entrenched into our lives. This is why, for a lot of you, this next bit might hurt a little. So entrenched is this myth I am going to talk about that the instinctive brain will not be able to cope with the challenge. He will shut down, stick his fingers in his imaginary ears, and shout "la la la." You may stop reading, which is a shame because you will miss out on some tremendous swearing later on, will never meet Science Columbo, and will not understand the joys of Paltrow Science.

PUBLIC ENEMY NUMBER ONE

The myth that I am going to talk about is firmly ingrained within society and the media; is frequently endorsed by celebrities; has dozens of books published about it every year; and has many dedicated websites and forums, products available for sale, and whole sections dedicated to it in health food shops, pharmacies, and supermarkets. It is a ritual undertaken by many millions of people all around the world, producing countless ringing anecdotal endorsements. It is impossible to avoid, it is believed by almost everyone, and yet it is completely ridiculous. So, like pulling off a Band-Aid, I am going to do this quickly to get the pain out of the way. Are you ready? 3, 2, 1 . . . Detox isn't real.

Right, that wasn't so bad, was it? The concept that we can detoxify our bodies by controlling our dietary intake is benchmark pseudoscientific bullshit. In terms of our biology, it makes absolutely no sense at all, it has no basis in fact, and there is virtually no evidence that its effects are real. Yet it supports an industry worth billions of dollars every year, has many powerful and influential advocates, and frequently makes bright and educated people fall under its spell. It is one of the biggest cons being sold to modern society, offering false promises of health and pseudoscientific justifications for its effectiveness. It is based on nothing more than flimsy fragments of anecdotal evidence but uses them to extract huge amounts of money from often-vulnerable people. It creates unpleasant rules, fears, and negative associations with food and modernity and has a nasty habit of offering dangerous and irresponsible advice. I fucking hate the word *detox* and detest the industry that has grown to exist around it.

THE CIRCLE OF BULLSHIT

Actually, I may have slightly misled you. In one respect detox is real. *Detoxification* means ridding your body of something that has poisoned it. So, if you are unlucky enough to suffer from alcoholism or other substance abuse, there is a chance you might undergo a period of detox as part of your treatment. Similarly, if you are unfortunate enough to poison yourself and you are rushed to hospital in crippling pain, bleeding from your stomach and eyes, then you may well have to undergo a medically administered detox procedure. But one thing is for sure: That procedure will not be a green juice diet. It will not involve lemon water and cayenne pepper. It will not be a specially formulated detox tea followed by a cleansing foot rub.

The detox myth is threefold. First is the myth that modern life constantly exposes us to unprecedented levels of dangerous toxins. Second, we are led to believe that our bodies are not capable of removing these toxins and they remain with us, stored somehow within our tissues, causing numerous problems and diseases. Third is the myth that certain foods, therapies, and treatments remove these toxins. In a remarkable feat of pseudoscientific circularity, all three are told simultaneously, all three are dependent upon one another, and all are encapsulated in a simple word. *Detox.* A glorious signpost for our gullibility. An industry built out of nothing. An enormous monolith constructed on the flimsiest of foundations.

It is time to pull apart these three detox myths and look at some of the reasons why they are so ubiquitously held.

THE MYTHICAL TOXICITY OF MODERN LIFE

You cannot have a "detox" without some sort of "tox," and although for some people detox regimes are undertaken in search of weight loss or in response to periods of conspicuous indulgence, many do so in the mistaken belief that modern life (and especially urban life) leads to us being inundated with levels of toxins that our bodies are not evolutionarily adapted to. Here are a few examples of how this idea is communicated:*

From "Medical Medium" Anthony William on Gwyneth Paltrow's website Goop: In this modern era, we are bombarded by toxins of every kind imaginable. Our bodies are subjected to an onslaught of dangerous chemicals on a daily basis from things like air pollution, plastics and industrial cleaning agents, not to mention the thousands of new chemicals introduced into our environment every year. Toxins also saturate our water reservoirs, fall down from the sky, and hide out in our homes and workplaces. This has become an unfortunate reality of modern life.

From Michelle Carlson, fitness instructor: A diet full of processed, high-sugar and high-fat foods (the typical busy American diet) can leave behind metabolic waste products in the system and even interrupt the normal hormonal fluctuations of the body. This, in turn, can make for a breeding ground for illness.

From shape.com's review of the Top Ten Diet Cleanses of 2014: Detoxing—or removing unhealthy toxins from the body—is one of the main reasons people give for wanting to do a cleanse. Toxic overload can make you feel sluggish, lead to acne, and can cause allergic reactions—among a host of other ills.

* Any bloggers' websites are quoted from the autumn of 2016, when I wrote the book. I say this because sometimes health claims go missing after I point them out on the Angry Chef blog.

The dangerous toxicity of modern life is commonly held up as causing the many health problems associated with modernity. And yet these mysterious toxins are usually very poorly defined. We are left in no doubt that we are under attack, and yet the source and nature of these attackers is rarely discussed. All we know is that there are bad chemicals everywhere doing us great harm, a blight on our modern lives.

It is true that the human body is exposed to literally millions of different chemicals every single day. But everything is made of chemicals. There is a tendency for detox advocates to separate these chemicals into good and bad categories, particularly along the lines of natural = good and unnatural = bad. We shall examine this strange fallacy in chapter 13 on convenience foods, but it is important to remember when considering toxicity that it is the dose that makes the poison. Water, a commonly ingested chemical, will kill you if you consume enough of it. And botulism toxin is an entirely natural substance (you could probably make it organically if you wanted to), yet one of the most poisonous that we know of.

There is quite simply no evidence that our bodies are subject to an unprecedented onslaught of toxins, and no evidence that we are being harmed by modern life. In fact, we are healthier than we have ever been: We live longer, contract fewer diseases, and have a food and water supply with lower levels of dangerous contamination than at any point in human history.

"YOU ARE NOTHING BUT A FESTERING VESSEL FULL OF EVIL TOXINS"

So, we have found our villain: toxins. The second idea is that these vicious unnamed toxins build up in our bodies and are not removed by natural processes. Here are some fairly typical examples (trust me, it is easy enough to find your own):

From Anthony William on Goop: Most of us are carrying around heavy metals that have been with us for almost our whole lives and which have burrowed deep inside our tissues. Unfortunately, it is these "old" metals, the ones that have been lurking in our system for prolonged periods of time, that pose the greatest threat. For example, over time toxic heavy metals can oxidize, causing damage to surrounding tissue and promoting inflammation. They literally poison our bodies, and can inflict damage on virtually every system and organ, including our brain, liver, digestive system, and other parts of our nervous system. Toxic heavy metals put an immense burden on our immune system, leaving us vulnerable to a variety of illnesses.

From NPR in a feature on celebrity detox guru Dr. Alejandro Junger: One of Junger's fundamental arguments is that our body is full of toxins we've picked up from food and the environment. These toxins slow us down and make us sick. As Junger recently tweeted, a "main cause of dysfunction in the body is the presence of obstacles [toxins] to the normal functioning of things."

In the first passage, there is talk of heavy metal toxins, so it seems fair to investigate this a little. The advice in that particular excerpt comes from Anthony William, the "Medical Medium," who gets his information from a spirit-world guide.

Heavy metals toxicity is a real thing, particularly in the case of lead, which is of huge potential harm to developing infants. Since the widespread removal of lead from gasoline, pipes, and paint a number of years ago, there is little evidence in modern developed countries of dangerous levels of lead poisoning affecting the population's health. Another potentially dangerous source could be arsenic, and although there are some areas of the world where this can naturally reach potentially harmful levels in drinking water, for most of us there is unlikely to be any risk in the amounts we consume.[1] Similarly, aluminum is everywhere, in our food, drinks, and even the air we breathe, yet the average amount we all consume is not thought to be in any way dangerous.[2] Whichever metal we look at, there is simply no evidence in the nonspirit world that the level of exposure we experience every day has any detrimental effect on our health.

In this statement lies something key to the detox myth. It is true that we are exposed to hundreds of potential toxins every day, that we ingest them, that they are in our foods, and that they contaminate our water supply. The world in which we live has always been a seething ball of chemicals and, throughout the history of life, exposure to potentially harmful substances has always occurred. For many this truth is uncomfortable, largely because our instinctive brain likes things to be black and white. The idea of degrees of toxicity does not sit well.

When science tells us "we are all exposed to poisonous heavy metals, but the level of exposure is not currently thought to pose any dangers to human health," all our instinctive brain hears is "Exposed! Poisonous! Danger! Health!" and he starts to run around in panicked little circles. It does not help that there are examples of industrial chemicals causing harm to humans. In the case of certain substances there is even some truth in the potential for them to build up in the body. Certain persistent organic pollutants (POPs) can accumulate in adipose tissue (fat) and have been linked to a number of chronic diseases.[3] The industrial use of these chemicals has been almost eliminated and they have been gradually decreasing in the environment, but unfortunately small amounts still persist.

It is thought there are POPs present in almost everything we eat, and yet in many ways it is more a testament to the technology that allows us to detect these chemicals in vanishingly small amounts than a sign that we are under attack. Our instinctive brains are just not adapted for this modern way of viewing the world, for the mind-boggling investigative power of mass spectrometers and high-performance liquid chromatography (HPLC) analysis. The easier it is to analyze and deconstruct the chemical composition of the environment, the more traces of chemicals you are going to find.

AN IMAGINARY CURE

Q: How do you deal with a made-up problem? **A:** A made-up solution.

As I have explained, the idea that our food and environment is full of potentially unprecedented levels of damaging toxins does have a grain of truth running through it. There is an even smaller thread of truth running through the idea that these mysterious toxins are building up in our bodies, but this is largely confined to a small group of banned chemicals, currently present at harmless levels. When we get to the third great detox myth, any connection with reality is severed completely. It appears that we are so convinced by the detox myth that we have stopped looking for evidence. Here is another selection of quotes, all from various lifestyle and detox blogs:

> **From Anthony William on Goop:** Wild blueberries (only from Maine): Draw heavy metals out of your brain tissue, healing and repairing any gaps created by oxidation when the heavy metals are removed. It is important to use wild blueberries, as they possess unique phytonutrients with special detoxifying capabilities.

> **From health blogger Madeleine Shaw:** Grapefruits: Containing natural vitamin C and antioxidants, these are both amazing for giving your liver a good clean.

From Doctor Oz on his kale, pineapple, and ginger detox drink (juicer required): This purifying beverage contains kale to cleanse the kidneys . . . .

"Kale to cleanse the kidneys," one statement that encapsulates the insanity of detox claims. Just to be clear, there is no evidence that any of these foods have any of the effects mentioned. None of them removes toxins and none of them helps our bodies to do so. None of them cleanses the liver or kidneys, whatever that might mean. I generally pride myself on an ability to pick apart the misunderstandings and confusions that result in false beliefs being born, but in the case of detox diets and detox products, there is no mystery to pick apart. Despite it being a multibillion-dollar industry, no clinical study has shown any commercially available detox diet or treatment to have any clinical effect in the removal of toxins.

The good news is that our bodies have excellent systems for removing any potential toxins. Our livers and kidneys evolved specially for this process, and unless you have a fairly fundamental problem with them, they will never need any help. Similarly, our skin, lungs, and digestive systems all play a role in eliminating toxins from our body. Even if some toxins did remain, there is very little evidence that any food would help get rid of them.

Cilantro is often quoted as being a "miracle" detoxing ingredient, and this is presumably based on the limited effects seen in two animal studies, one on cadmium in rainbow trout and one on lead in contaminated mice.[4] Cilantro is about the most studied of all foodstuffs when it comes to detoxification and yet there are no human studies showing any effects and no studies showing anything other than a small impact on heavily poisoned animals. As for blueberries (but remember, only the ones grown in Maine), ginger, kale, walnuts, garlic, green tea, and numerous other substances named as detoxifying, there is no evidence at all. That is not to say they are not good for you, just that they cannot unpoison you, especially when you haven't been poisoned.

If you still don't believe me, I would suggest asking anyone who is trying to sell you a detox product which toxins they are claiming to remove. It is a fairly simple question and should elicit a fairly simple answer. If they answer that it is a variety of different toxic substances, ask them to name just one. Once they have done this, ask them if there is any specific evidence of an increase in the levels of excretion of this particular toxin after undergoing the treatment. If you are being detoxed, the toxin will have to be excreted somewhere and that could easily be measured. Unlike a lot of areas of nutritional science, the detoxing effects of different foods or products could be fairly easily studied. Experiments could easily be designed to prove any treatment has a real effect, and if those experiments have not been done, we have to ask why. Unless you are a rainbow trout concerned about cadmium levels, there will not be any useful, applicable evidence at all.

What is going on?

Pseudoscience researcher and writer Emily-Rose Eastop, one of the founders of the excellently named I Fucking Hate Pseudoscience (IFHP) website, a popular online science advocacy page, suggests that people can be quite easily led when it comes to these terms:

> The problem is that although believing in, say, detox or raw-foodism may (at least for those who don't become fixated on it) be pretty much harmless in isolation, it and other "harmless" forms of pseudoscience can act as a gateway to other, more directly dangerous ones. In order to believe in the efficacy of detoxing, you have to reject the multiple lines of overwhelming scientific evidence against

it, or at least not require evidence as a proviso for taking on new beliefs about physical things like the body in the first place. Without this proviso, one can potentially believe anything. In order to believe, otherwise reasonable people have to abandon the tenets of reason.

Paul Rozin, professor of psychology at the University of Pennsylvania, has researched people's beliefs about food and thinks that we are likely to be susceptible to the wild claims of detox regimes over the more measured and sensible dietary advice being offered by official sources.

People like the idea that there is one bad thing that they can just get rid of. People don't want advice that tells them they can make changes and improve a little. They want simple rules about good versus bad foods. They want to be told about huge improvements and even though they might know it's unlikely to work, they are willing to fool themselves time after time.

For many people, there is more than just the potential for health improvements. For some, to detox is to achieve an inner purity and to signal your virtue to the world. In a revealing passage discussing a detox smoothie recipe, the clean-eating blogger Ella Woodward comments:

A glass of this will set you up perfectly for your day, and you'll feel fantastic. Not only because of all the goodness and nutrition that you're drinking, but also because there's always a sense of pride and happiness associated with knowing that you're doing what you can to look and feel your best. I think consciously looking after your body really instills an awesome sense of pride and self-worth, which everyone needs a little extra of most of the time.

Helen West is a registered dietitian (diet scientist) who writes about myths and misunderstandings in food on her Food and Nonsense blog. For her, much of the appeal is quite clear.

Like a lot of this stuff, going on a detox actually provides a more acceptable euphemism for wanting to get thin. The word is often not used to refer to flushing out toxins, but is just about extreme calorie restriction. It is used in a slang way, without thinking. Most people do not believe they are actually flushing toxins from their body.

For many this may well be true, and we will look at the way nutritional pseudoscience is used to disguise weight-loss goals later on, but within the teaching of less legitimate nutritional sciences, the detox myth is taken quite literally. In a 2011 survey for the *Journal of Alternative and Complementary Medicine*, 75 percent of naturopaths reported prescribing diet-based detox measures to "treat" health issues.[5] Health writer Ian Marber, who trained in nutritional therapy at the Institute of Optimum Nutrition, told me:

You are taught that an apple has been sprayed twenty-two times before it gets to you. Although how this affects the human body is not explained, the number is used for dramatic effect. You are told that the human body cannot cope with the toxins of modern life, that it places a stress on the human body, and that it overwhelms the liver, pushing it over its natural capacity. You are also taught how natural substances can help your liver get rid of these stored toxins. It is very much the selling of fear.

Fear is effective because it taps into something deep-seated and primal. We fear a strange and invisible enemy, perhaps trying to label our discomfort at the seemingly random progress of our lives and our health. People have always believed in the purity of times past and the contamination of modernity, with a pure, unpolluted paradise at the origin of most religions. As each of us age, we have the tendency to mistake our own decline with that of the world, to believe that there is something good and pure about the past that we have lost, when what we really mourn is the loss of our own youthful vitality. The old will always try to damn the present as it exists for the young, and there is no better way to damn something than to declare it unclean.

The big question

For me personally, an early engagement with the world of detox opened up a hugely perplexing question. A couple of years ago a friend had returned from five days at Glastonbury Festival (the Coachella of the UK, but bigger) and by all accounts had fully embraced the spirit of the festival. She is intelligent, with a degree in food science and fifteen years' experience working in the food industry. She decided that she needed to rid her body of toxins and, using advice from a popular detox book, she went on a cleansing program that involved seven days of consuming only lemon water. After a week of pain, hunger, and hallucinations, with a body desperate for some sort of nutrition, she ate a family-size bar of chocolate in a few seconds, nearly passed out, and had to call an ambulance when a migraine made her think she was having an aneurysm.

This is a nice illustration of how ridiculous and damaging detox can be, but the part that perplexes me the most is this: How is it that someone bright, intelligent, and informed is still susceptible to the pseudoscientific spell of detox? If I had asked her about the Loch Ness monster, 9/11 conspiracies, fake moon landings, homeopathy, or antivaccine campaigners, her opinions would have been thoroughly sensible. But when it comes to imaginary conspiracy toxins and magical superfoods, despite her qualifications and experience in this area she was locked into accepting bizarre claims. So locked in that she was actually willing to risk her health because of it. When it comes to diet, why do otherwise sensible people hold such ridiculous beliefs?

Whatever I might think about it, there is one really good thing about detox. For our simple, easily led instinctive brains, it is a wonderful signpost for spotting pseudoscientific bullshittery. In fact, it is **Rule Number 2 in the Angry Chef's Guide to Spotting Bullshit in the World of Food**: *Everyone trying to sell you a detox is peddling a myth.*

Chapter Three

THE ALKALINE DIET

Isaac Asimov once said that "creationists make it sound like a 'theory' is something you dreamt up after being out drunk all night." There is an understandable confusion in some people's minds, because the word *theory* has slightly different uses in normal conversation and in science. After one particularly eventful night out during my misspent youth, I once came up with a theory that cats were capable of controlling people's minds. This turned out not to be the case (I hope), but in the world of science it would have been wrong to refer to it as a "theory." I should have said that I had a "hypothesis," a hypothesis being an idea that has yet to be proven. Only when you have substantial supporting evidence does something move to the next stage in the scientific method and become a theory, meaning that it is accepted as a valid explanation of observed phenomena. Within science there are rarely competing theories, just competing hypotheses looking for evidence. This may seem like splitting hairs, but it is worth keeping in mind for this chapter. When someone says they have a theory that challenges the conventions of science, they probably should not be taken too seriously.

WHY DOES ELLE MACPHERSON PEE ONTO A STRIP OF PAPER?

I struggled through a degree in biochemistry twenty years ago and forgot most of it, but I am occasionally asked science-related questions. A while ago someone asked me why lemons are alkaline, which is a slightly odd question for anyone with even a basic grasp of chemistry (or anyone who has ever tasted a lemon). Lemons are strongly acidic, as are many fruits, which gives them their sharp, acidic taste. I was curious as to why anyone would think the opposite, so I decided to do a little digging. Underlying this simple question was a bizarre world, one that I was surprised to learn has developed huge power and influence.

For some reason, the alkaline diet (sometimes called "the alkaline ash diet") has become the regime of choice for many celebrities. Among those believed to follow its principles are Victoria Beckham, Gwyneth Paltrow, Robbie Williams, Jennifer Aniston, and Miranda Kerr. Supermodel Elle Macpherson is rumored to carry around a pH testing strip in her bag to test her urine (more on that later).

Such endorsement by glamorous celebrities with their enviable, age-defying looks and seemingly perfect lives certainly gives the diet public legitimacy. It regularly features in newspapers, fashion titles, and lifestyle magazines, with a number of health bloggers and "nutritional therapists" explaining the principles and providing helpful recipes that embrace the alkaline-eating philosophy.

So, what is alkaline eating?

OK. Back to school time. pH is a measure of the concentration of hydrogen ions in a solution. The more hydrogen ions there are, the more acidic the solution. The pH scale ranges from 0 to 14, with a pH of 7 being neutral, anything below 7 being acidic, and anything above 7 being alkaline. A pH of 1 is highly acidic; a pH of 13, strongly alkaline. I appreciate that for many of you this is about where education in chemistry stopped, but don't worry; the same seems to be true for many of the diet's advocates.

The alkaline diet is based on the premise that our bodies like to exist in an alkaline state and different foods will either acidify or alkalize us when consumed. To maximize our health, we should focus our diet on alkalizing foods, and cut back on, or avoid, acidifying ones. Curiously, some foods that are actually acidic (such as lemons) are thought to have an alkalizing effect on the body and some foods that are alkaline (such as dairy) are thought to have an acidifying effect.

This is a slightly odd belief, but there is at least some evidence to back it up. In the early part of the twentieth century some pioneers of nutrition research analyzed foods by looking at what remained after burning them in a calorimeter (a device where foods are burned to see how many calories they contain). The resulting ash was dissolved in water and its pH measured, hence the "ash" part of the diet's name. In 1912 an equation was proposed to enable the excess acidity or alkalinity of foods to be determined from their nutritional composition.[1] Unfortunately, this early work contains some slightly ambiguous and confusing terminology, which would produce a number of misunderstandings many years later. In summary, all phosphate and sulfate ions are considered acidic and all calcium, potassium, sodium, and magnesium ions are considered alkaline.

In terms of the alkaline diet today, using this formula, foods are categorized neatly into acid or alkaline forming. It creates simple rules that our instinctive brains can easily follow, and strongly advocates for the consumption of plenty of fruits and vegetables, thought by most people to be a good thing. Although there is some debate and inconsistency, foods thought to be acid forming include, in general: dairy, meat, fish, beans, sugar, coffee, lentils, rice, potatoes, wheat, bananas, cherries, oils, fats, and most seeds and nuts. Foods thought to be alkaline forming include most fruits and vegetables and . . . well, not much else, to be honest. It is perhaps not surprising that sticking to these foods is likely to cause weight loss, because there's not much you are allowed to eat.

The only real problem with the "theories" underlying the alkaline diet is that they are a huge steaming pile of imaginary bullshit. Although the early research into the ash of foods has some vague validity, the jump to any implications about the health-giving nature has no basis in reasoned science at all. It is all imagined and ridiculous, and when we look at where this strange "science" has come from, it is clear not only how crazy it is, but how it hides a dark and festering heart.

WHY THE ALKALINE DIET IS SUCH A LOAD OF SHIT

The human body is a miraculous thing, and within its many structures and functions, a number of different pHs are maintained. The stomach is highly acidic, with a pH of 1.5 to 3.5, which helps it break down foods as they reach it. The skin is quite acidic, too, helping to protect against bacteria and infection. Our blood is slightly alkaline, maintained at a pH of somewhere between 7.35 and 7.45, and as a number of vital processes depend on it, it is very important for our bodies to maintain this level at all times. In this limited way, the advocates of the alkaline diet are correct: When it comes to our blood pH, a slightly alkaline state is preferable.

Where the alkaline hypothesis falls down, however, is in the belief that food intake will somehow alter blood pH. It is true that in a single study (much cited by the diet's followers), the effect of certain foods has been seen to alter blood pH, only it did so by tiny amounts (0.014 pH units), well within the normal range.[2] If our blood pH shifts away from the optimum pH range at all, a number of processes quickly get to work to maintain it, most important, respiration. When we breathe out carbon dioxide, it raises the pH of our blood, meaning that altering our rate of breathing is a far more significant control mechanism than diet. There are also regulation systems in our blood and kidneys, which means that we quickly excrete excess acidic substances should our blood pH become too low.

The consequences of our blood pH changing even slightly are potentially very serious. Alkalosis (when the blood pH becomes uncontrollably alkaline) starts with confusion, tremors, muscle spasms, and vomiting and very quickly leads to coma and death. If the balance of our blood pH depended upon our food intake, we would all die very quickly.

Even the construction of any list of acid-forming foods is riddled with confusion and misunderstanding. The original formula by which the "acid load" of foods is calculated has a number of technical and classification problems associated with it (for chemistry fans who cannot be bothered to look up the reference, anions and cations are wrongly classified as acids and bases), leading to many items being wrongly labeled as acidic.[3] It is also generally thought of as being far too simplistic, ignoring the potential for many competing metabolic effects in our body that might interfere with acid generation. For instance, more recent studies on milk have indicated that it decreases the acid load, yet this is still firmly classified as "acidic" within the alkaline diet literature.[4]

Central to many health claims of the alkaline diet *hypothesis* is that phosphate (alleged to be "acidic") causes problems for our retention of calcium, leading to potential problems for bone health, but these claims are at odds with more recent evidence, which suggests that phosphate actually has the opposite effect.[5] This new evidence shows that phosphate should be excluded from the calculations when assessing the acid loads of foods, but perhaps unsurprisingly, the diet's followers prefer to stick to a much-disputed hundred-year-old formula. If they did exclude phosphate, all dairy and grains would fall into the alkaline category, making it a very different diet indeed.

Often followers are told to test the pH of their urine as an indicator of the diet's effectiveness, and although this is a good test as to whether or not your kidneys are working, it will not tell you anything about your blood pH. When Elle Macpherson next pees on a strip of paper, she will be gathering useful information about what she is excreting and the state of her kidney function, but little else.

Perhaps the most puzzling thing about the alkaline diet is the source of all the nonsense about its miraculous health benefits. The original literature focuses only on the pH of calorimeter ash and says nothing of any other benefits, yet the diet's followers talk of remarkable improvements, dramatic weight loss, increased vitality, increased immunity to disease, and, uncomfortably, the prevention and treatment of cancer. It is maybe not surprising that the source of these claims is rarely discussed, because whereas the alkaline ash hypothesis of foods is just an oversimplification of some complex chemistry, the reasons for its health claims are beyond crazy.

The curious case of Robert O. Young

Pretty much all health claims relating to the alkaline diet can be traced back to the "groundbreaking research" of a certain Robert O. Young, an American naturopath and author of a number of books, including *The pH Miracle*, in which he outlines his "theory" of a new biology. It is largely inspired by the work of Antoine

Béchamp, a nineteenth-century French scientist who believed in the concept of pleomorphism, the idea that matter can take many different forms. Béchamp and Louis Pasteur were researching the causes of disease around the same time, and Béchamp's hypothesis directly conflicted with Pasteur's germ theory of disease. Béchamp believed that the germs that Pasteur had observed were actually symptoms rather than causes of illness, produced by the body as a reaction to a diseased state. As time progressed and more and more evidence supported Pasteur's work, Béchamp's idea was eventually discounted. Pasteur's germ theory won the day and went on to revolutionize the health outcomes of everyone on the planet.

Although the idea that germs are produced by the body in response to disease seems crazy to us now, Béchamp's hypothesis should *not* be classified as dangerous quackery. Given the state of knowledge at the time, the idea that germs were produced by the body was plausible, despite being wrong. We should credit Béchamp with adding to the sort of debate that moves science forward. In the search for truth, science needs to work hard to develop as many different hypotheses as possible, then search for evidence to confirm where the truth lies.

For most of us Antoine Béchamp and pleomorphism are little more than amusing footnotes in the annals of science. Not for Robert O. Young, though. His beliefs center on the idea that Béchamp was correct all along and also that all disease is caused by the body becoming acidic. This is not an exaggeration. In *The pH Miracle*, Young claims that the "overacidification of body fluids and tissues underlies *all* disease." Everything from the common cold, hepatitis, AIDS, allergies, diabetes, and influenza is said to be caused by our bodies becoming acidic. Not only does he claim that he has seen his diet increase vitality, stamina, and mental ability, he also says he has seen people cured of type 1 diabetes and cancer simply by switching to alkaline foods. Given that the acidification of body fluids does not occur and his classification of foods is based on an oversimplification of some hundred-year-old chemistry experiments, these are impressive claims.

But there is more! Robert Young also claims to have seen human red blood cells transform into bacterial cells in an acid environment and believes that the genetic material of microorganisms changes depending on their diet. He goes on to explain how we should drink four liters of water every day—but not just any water. It has to be distilled, because tap water is full of toxins and bottled water is dead. Yes, I did say "dead." Apparently water can be biologically active and "alive" in some unexplained way, but if it tragically dies, you can enliven it by adding hydrogen peroxide (yes, bleach) to make it more alkaline. Even better, if you add some acidic lemon juice to the hydrogen peroxide water, you will make it even more alive and even more alkaline (because the acidic lemon is of course alkaline), which is such a bizarre claim that it actually makes my head hurt.

I am hoping that most of you will be able to spot some of the problems with Young's ideas. His theories are not theories in the scientific sense. It is stretching the point to call them a hypothesis. To claim that germs do not cause disease, you have to ignore nearly 150 years of scientific progress, during which death rates from infection have plummeted. If there were even a grain of truth in the idea that microorganisms can transform and change their genetic material, we would also have to start again with biology, a field that depends on the idea that genes are inherited. If water is somehow alive or dead, then all chemistry and physics would be wrong, too. So, if Young is correct, and the alkaline diet is to be believed, then all of science is wrong.

Followers of the alkaline diet need to understand that in accepting the diet's philosophy they are rejecting the whole of mainstream science. Perhaps unknowingly, every celebrity advocate is buying into pseudoscientific bullshit of the highest order. Although some people might say that the likes of Robert Young and other alkaline theory advocates are just trying to get people to eat a few more vegetables, the unavoidable reality is much, much darker.

Kim Tinkham is perhaps the most famous victim of this blind acceptance. A patient of Robert Young's, Kim appeared on Oprah Winfrey's TV show having refused surgery for stage 3 breast cancer, convinced that she could heal herself through her own mind and her diet. She gave a number of glowing testimonials for Young's alkaline plan before succumbing to her disease, dying sometime after a press release had been issued by Young declaring her as cancer free. Although the exact details will never be known, it is hard to think that Young's mistaken beliefs did not play a part in her tragic death.

There is perhaps a happy footnote to Young's story and one that surprisingly came to light recently. In 2016 he was convicted of theft and practicing medicine without a license, and at the time of this writing he has pleaded guilty and is serving time in jail. I do hope he enjoys the food.

Please make it go away

So, Robert Young has been convicted, and his ideas about an alkaline diet go against pretty much all of science. And yet like some unstoppable virus they continue, with advocates of his principles happy to spread this ridiculous and even dangerous doctrine. In a recent newspaper interview celebrity nutritionist Natasha Corrett was quoted as saying, "The body can't get cancer in an alkaline state; cancer creates disease in the body through acidity. . . . Someone who eats acid-forming foods is more likely to get common colds, bad skin, and dull hair." In another interview Natasha said,

"When your body is alkaline you have clear skin, greater concentration, shiny hair—all the things you want with the bonus of weight loss."*

Clearly Natasha is more worried about dull hair than I am, but these claims could prevent someone from seeking medical treatment in the face of a life-threatening illness. What is worse, it is all based on the ramblings of a convicted criminal. Why on earth has the alkaline diet not gone away?

As with detox, one reason may well be the hidden weight-loss goals that lie at the heart of the wellness industry. Along with a huge list of restricted "acidic" foods, for followers of the diet there are also a number of complex rules about combining foods, leading to further food fear and calorie restrictions. Although the science is complete nonsense, anyone following the diet would be likely to lose a lot of weight. Restricted foods are pretty much anything containing significant amounts of carbohydrate, fat, or protein, which leaves some fruits and most vegetables, all washed down with delicious bleach water. It's a long way from being healthy or balanced, but you will most likely shed some pounds.

The alkaline diet is based on simple rules with the sort of emotive language likely to result in people sticking to them. Our minds associate anything acidic with being bad, corrosive, and dangerous. Acid rain, the acidification of the oceans—they are signs of modern decline, the human desolation of the natural world, the destructive power of modernity. Anyone who has ever had much contact with strongly alkaline substances will know that they are equally damaging, but the use of language makes them seem far more natural and benign than their dangerous opposite. Foods

* When asked to clarify the claim that "the body can't get cancer in an alkaline state," according to the interviewer, Natasha responded that "it's impossible for a body to be alkaline enough to resist cancer, but [she settled] for the less contentious idea that it's mainly about keeping the body in its 'optimum' state."

are neatly classified based on vaguely sciencey-sounding terms, terms we all remember from school chemistry and that make us feel like we understand something new about the world.

Ian Marber has long decried the alkaline diet myth, calling it the "benchmark of quackery." Concerning advocates of the diet, he believes that it tends to attract a certain type:

> Alkaline seems to attract the same sort of people and they are often those lacking in talents elsewhere. There is a low barrier to entry for this sort of nutritionist and you will be told throughout the training that we are all special, we are all individual. There is often a strong sense of entitlement, a sense of narcissism, where everything is about the practitioner and how they feel. They think, *if it worked for me, it must work for everyone*—and for people who are only interested in weight loss it does work. Any food plan that produces rules will make you lose weight, but they mistake that for the science being somehow valid.

The alkaline diet also strongly taps into the idea that health outcomes are the fault of the individual and that everything is down to personal choice. This is also a strong theme throughout nutritional pseudoscience, and naturopathic medicine in general. In the words of Robert Young, "If you get sick, it is your own fault and not the cause of some phantom virus that you can blame to cover your own lifestyle and dietary transgressions."

So, cancer, allergies, type 1 diabetes, mental health issues, obesity, and the common cold are all neatly blamed on the individual, all wrapped up in a neat package of self-loathing, designed to inspire us to change our behavior. Our instinctive brains intensely dislike the random nature of illness and suffering, the idea that shit just happens sometimes, that good people die, that cancer might strike any of us without warning. It does not fit with our desire to make sense of the world. If you can frame everything in terms of simple black-and-white choices and create a hidden secret that makes illusory sense of frightening randomness, then a compelling proposition is created. The alkaline diet creates a false illusion of control in a random world, and this is so powerful that we can become willing to discount reason to believe.

And that leads us to **Rule Number 3 in the Angry Chef's Guide to Spotting Bullshit in the World of Food**: *They will always tell you it's your own fault.*

Chapter Four

REGRESSION TO THE MEAN

*The art of medicine consists of amusing the patient
while nature cures the disease.*

—VOLTAIRE

THE TROUBLE WITH TOMMY

Many years ago I used to have this little commis chef called Tommy. Angry little Mancunian boy he was, full of nineteen-year-old rage and brash self-confidence. He wasn't a bad chef, but neither was he a future Anthony Bourdain. During a busy service, especially when things were really hitting the fan, he would either be a complete star, or a total nightmare. I think this probably had something to do with his nocturnal activities after work. Tommy was a good-looking guy and this was Manchester in the mid-1990s— there was plenty of nightlife around.

The thing that used to frustrate me the most about Tommy was this: When he had done really well, and at his best he could really cook, I would take him aside after service, buy him a beer, and spend a good ten minutes telling him how great he was. I'd say he had a real future, that he could run a kitchen of his own in a few years' time. Sure enough, whenever I did this, the next service he would take a couple of steps back toward his space-cadet self.

When he had done badly, and he could really mess up a service sometimes, I would give him a kicking (a metaphorical one; I was never one of those chefs). The next time out he would pull himself together and we would get a decent few days out of him.

Me and Tricky (my old second chef—a bit of a character) would often talk about this, trying to work out how we could get the best out of Tommy. It was starting to become a real problem, because although we knew he had talent, we could never rely on him when we were planning rotations as we didn't know which Tommy was going to turn up. Tricky was always a bit more pragmatic than I was and frequently told me to "stop being so nice to the little shit." According to Tricky, it just made him overconfident and complacent. Tricky would tell me that the key to managing chefs like Tommy was a constant level of admonishment and fear. "Just give him a bollocking [a good talking-to, for you Americans] every time, whatever he does. That way he'll always get better." Often, when I was not around, Tricky would do just that.

Suffice it to say Tommy and Tricky never really got along and eventually, on one memorable evening, I had to stand between Tricky and Tommy's slightly terrifying father who came into the restaurant to sort us out after Tommy had called him in tears. I heard later that Tommy had ended up working in a call center, where I doubt they had the same standard of admonishing words, so we never did get the chance to fully test out Tricky's theory.

If my second chef had been the behavioral economist Daniel Kahneman, things might have been a little different. I doubt Kahneman would have been quite as good on the stove, but he may have been able to provide a slightly better insight into what was going on with Tommy. He once observed a similar phenomenon in fighter-pilot instructors while working as a psychiatrist for the Israeli Air Force. Kahneman was told by the instructors that when they praised trainee pilots after a good flight, their performance would always drop away next time, but when trainees were punished after a poor run, performance always improved. The conclusion of the instructors, much as in Tricky's case, was that a good old-fashioned talking-to is the best form of motivation. Daniel Kahneman, however, saw it differently. He knew from previous studies that by far the most effective motivational strategy was encouragement and reward. So, what was going on? Why did my little ball of Mancunian attitude and a group of trainee fighter pilots seem to respond so differently to the predictions of psychiatry?

The answer is quite simple really, not beyond the comprehension of me and Tricky over a beer, but not within our sight either. The reason Daniel Kahneman has a Nobel Prize and I have a certificate for Second Best Shortbread in the Long Marston Village Show (I was robbed) is that Kahneman spotted the phenomenon of *regression to the mean* affecting performance, whereas Tricky and I just thought Tommy was an unpredictable little shit.

REGRESSION TO THE MEAN

Regression to the mean is a deceptively simple concept. The easiest explanation of it is to say that things even out over time. If you want to use slightly more academic parlance, you might state that "if a variable is extreme on one measurement, we would expect it to be closer to the mean (average) value on the second measurement."

Take my little Mancunian chef. His performance on a Saturday-night service was highly variable, and had different extremes. On occasion he could be brilliant, but sometimes he could be a huge liability. At other less memorable times, he was merely average, after which we would probably not pay much attention while getting a drink after service. If we take into account regression to the mean, it makes perfect sense that on the occasions when he was brilliant, as this is an extreme, he would be likely to regress back toward the average level next time. And on the occasions when he was terrible, also an extreme, he would also be likely to regress back toward the mean, in this case making an improvement. My pep talks and Tricky's talkings-to probably had little effect on his performance at all. Looking back, I do feel a little guilty for not noticing it at the time. I hope the call center worked out for him.

The interesting thing about regression effects is that although they make perfect sense, they are incredibly hard to spot. So hard, in fact, that regression to the mean was not properly categorized until the late nineteenth century, when one of the great minds of his day, Sir Francis Galton, observed it while studying changes in successive generations of sweet peas.

Let's just think about that for a moment. Regression to the mean entered scientific understanding after such discoveries as the universal law of gravitation, anesthesia, infrared radiation, electromagnetism, evolution by natural selection, and germ theory. It hides in plain sight, largely because our instinctive brains look for patterns and stories to explain away its effects. As I will try to show, if we can only learn to spot it in action, it can explain a great deal.

Regression to the mean in sport

Take the world of sport. Sporting activity is perfect for examining regression effects because it tends to involve a certain amount of skill combined with a degree of luck. In the United States there is something known as the *Sports Illustrated* curse. When a sportsperson is featured on the front cover of *Sports Illustrated* magazine, they will often suffer a subsequent loss of form. The same is true of many such accolades. The English Football League's "Player of the Month" will often become a spare part for the next few games. Athletes securing lucrative sponsorships from top brands frequently fail to deliver in the next big championship. Pundits and fans will be full of explanations for these failures. They will say that the athlete has become overconfident, that the opposition has found them out, or maybe they are feeling the pressure of expectation. But in many cases, the real cause is more likely to be regression to the mean. When athletes achieve great accolades, they are usually at the top of their game, and when you are at the top of anything, there is only one way you can go.

Even harder to spot is regression to the mean after a poor performance. When an athlete is suffering a dip in performance, at their lowest point, once again there is only one way things can go. Any nadir in performance is also likely to be the point at which the athlete will try to change something to help them improve. It might be new footwear, a new training technique, a new brand of putter, or a dietary alteration. If one of these changes is followed by a return to form, the athlete might attribute this change to the intervention they have made. The reality is probably just that a run of bad luck has ended, but try telling them that. Jamie, of gluten-free fame, started running faster after cutting out gluten, and now he is never eating gluten again.

REGRESSION TO THE MEAN AND THE RISE OF THE HEALTH BLOGGERS

So, what does this not-so-obscure, but difficult-to-spot phenomenon have to do with spotting bullshit in the world of food? Let's just say it's all around us, and may just give us insight into one of the great mysteries of the internet! Why do all the recent online insta-health-gurus have an uncannily similar biography? The template goes something like this:

*I was living my impossibly glamorous life as an **INSERT GLAMOROUS OCCUPATION HERE** at a hundred miles an hour, eating all sorts of junk and not caring what I put in my body. My health was really suffering. It was only when I started to* *take control of the food I was eating that my health improved. I started my **INSERT NAME OF MADE-UP DIET PLAN HERE** and it revolutionized my life. All my friends just begged me to share my recipes with them, and that's how my blog was born.*

Here are a few examples:

Vana "The Food Babe" Hari: My typical American diet landed me where that diet typically does, in a hospital. It was then, in the hospital bed more than ten years ago, that I decided to make health my number one priority. I used my new-found inspiration for living a healthy life to drive my energy into investigating what is really in our food, how is it grown and what chemicals are used in its production. I didn't go to nutrition school to learn this. I had to teach myself everything spending thousands of hours research- ing and talking to experts. . . . Most importantly, the more I learned and the more lessons I put into action, the better I felt and wanted to tell everyone about it!

From food blogger Ella Woodward: I started the blog as a way of dealing with a relatively rare illness, postural tachycar- dia syndrome, which I was diagnosed with in September 2011. The illness had a pretty devastating effect on my life . . . . So I decided it was time for something new and began researching holistic, natural healing approaches, which is how I started eating like this. Overnight I took up a whole foods, plant-based diet and gave up all meat, dairy, sugar, gluten, anything processed and all chemi- cals and additives . . . it's single-handedly the best thing I have ever done. Knowing that I'm giving my body the love and health that it needs is an incredible feeling.

Dr. Alejandro Junger, a Paltrow favorite and celebrity detox advocate, in an introduction on the blog mindbodygreen: He graduated from medical school in 1990, and moved to New York City for his postgraduate training. . . . His drastic change in lifestyle and diet from his move to New York City soon manifested as irritable bowel syndrome and depression. Becoming a patient of the system he was practicing was such a shock that it started a journey to search of an alternative solution to his health problems. His findings are the subject of his first book, *Clean*, in which Dr. Junger describes how he became aware of the toxicity of our planet.

From a *Daily Mail* interview with Natasha Corrett: Natasha . . . had struggled with "every single diet there is" for years, watching her weight bob up and down and never understanding why she always felt so tired and bloated. A chance visit to an Ayurvedic acupuncturist after injuring her back revealed her symptoms as "acidic," and she was put on a two-week eating plan of oily fish, green tea, whole grains, green vegetables and almonds. She cut out fizzy drinks (including soda water and tonic), sugar, sweeteners, peanuts, white flours and chocolate—and in under a year she lost two stone [28 pounds] and shrunk [sic] by two dress sizes.

Rest assured, if you scour the internet and research healthy-eating literature, you will find that every nutritional guru fits this profile almost exactly. They are well connected, leading an interesting, fast-paced life. They undergo a health crisis that conventional methods cannot treat. They heal themselves through diet and develop a new understanding through their personal journey.

They decide to share that new understanding with the world. This forms the basis for **Rule Number 4 in the Angry Chef's Guide to Spotting Bullshit in the World of Food**: *If someone fits this template, take their advice with a large pinch of Himalayan pink salt.*

What exactly is going on?

This phenomenon is curious. How can we explain why so many health and wellness bloggers have an almost identical format to their stories?

Here are a few potential explanations:

1. Health bloggers are actually cyborgs being created in a secret underground bunker, funded by the people of Big Avocado to drive sales. Unfortunately, the artificial intelligence (AI) software they are using can only handle a limited template.

2. Health bloggers are evil. They are cynically making up a story along proven lines and know that they will be believed because they seem likeable and attractive and project a lifestyle that many find aspirational.

3. Glamorous people living fast-paced and impossibly exciting lives are more likely to discover the secret to achieving health and wellness through diet than scientists working in laboratories.

4. Health bloggers are not very bright and falsely believe in the first thing they read when they Google "healthy eating."

5. Regression to the mean.

Much as I do enjoy the idea of an underground bunker, and this will feature as a plot line when they finally make the Angry Chef action movie, I am happy to discount option one.

A lot of people I know are inclined to believe option two, but personally I do not think that the sort of individuals who make a living communicating health advice are likely to be evil, however false or misguided their advice might be. Genuinely evil people could surely use their nefarious skills more effectively, perhaps with a career in bank robbery, cybercrime, or antivaccine campaigning. Or maybe they could just become lawyers.

I am assuming that if you have got this far into the book, you are happy for us to discount option three without much discussion. I would say that if you still find this option convincing, we should probably just call it a day. Maybe go and read a Dan Brown novel. I am not sure that this book is the place for you.

Option four is a possibility, but I am also not convinced that the kind of people who have successful careers, a good group of contacts, expensive educations, and are acutely media-savvy are likely to be completely witless. They may be poorly informed, and perhaps show little grasp of the importance of evidence, but I do not think any of them are especially dumb.

I tend to favor option five. I have a theory that much of the current plague of self-appointed internet health gurus can be explained by regression to the mean.

A theory? Don't you mean a hypothesis?
Well, maybe. Although to be honest, for it to be a proper hypothesis we would have to be able to design an experiment to test it.

What I meant to say was—I think that much of the current plague of self-appointed health gurus can be explained by regression to the mean. You don't have to be stupid or evil to hold some pretty strange beliefs.

Let me explain. When you are stuck in a difficult period of ill health, much like the athlete on a run of bad form, you will tend to look for solutions. This is especially true if conventional medicine does not have much to offer in terms of effective treatment. For many of the health bloggers we have discussed, the symptoms they were suffering were vague, poorly diagnosed, and have few proven treatment options. When modern medicine has little to offer, one of the few things people can control is their food. As their illness progresses and they start to feel worse, they will look for any intervention that might help. Many will find advice from alternative practitioners or the internet, often telling them to exclude certain things from their diet.

If someone has been feeling ill for a while, provided they don't have a chronic degenerative illness, then they are very likely to start feeling better soon. When someone is at their worst, this is the exact point at which they are most likely to make an intervention, and also the exact point at which their health is likely to start improving. And as we know, when something is improving, we like a nice, simple story. We look for evidence that can justify this newfound belief.

Things can escalate. Once we have an ingrained belief, in this case the belief that excluding a certain foodstuff causes a large improvement in health, if our health declines again (as it will if we have been feeling particularly good), we may well be inclined to exclude something else. Once again, if this happens at the nadir of the illness, the positive change and the intervention will occur at the same time. Another food exclusion will be linked to improved health and ingrained as a rule.

All that has happened is that stuff has evened out over time, but that story is boring, so our brain has created a dietary fad. And so false beliefs are born. Once we have seen something with our own eyes, we are extremely likely to accept it as the truth. We may hear the contrary testimony of a thousand experts, and we may read about a hundred well-conducted double-blind clinical trials, but nothing will shift a belief in something that we have witnessed for ourselves. Unfortunately, this is just the way the human brain is wired. It had evolved to believe its own expert witness above all else.

But why are they all so shiny?

Why is it, then, that all the successful health bloggers seem to come from glamorous, well-connected backgrounds? How come they all seem like ready-made new media stars, all photogenic and tech-savvy? Surely this is evidence that they are all part of a vile manipulative conspiracy?

Again, I think not. I am sure that for every new Instablogging star, there are a thousand other misguided clean eaters who have been through the same sort of life-changing dietary health discovery. It is only the media-savvy, photogenic ones with a good set of contacts that will be able to effectively project that discovery into the wider world.

I think that this can help explain why all the new food-health stars have near identical biographies. False beliefs in fad diets are caused by regression to the mean, the ability to command a high profile comes from preexisting contacts, and the bloggers' passion sadly comes from the way the human brain is wired.

And so it spreads . . .

As the media profile of our new star grows, more and more people will become aware of their message. Many looking for help will be suffering from health problems, often the sort of problems where conventional medicine has little to offer. At their own personal health nadir, a point where things can only get better, they may discover the blogger's new "INSERT NAME OF MADE-UP DIET

PLAN." Crucially, the exact details of the diet plan are irrelevant because as soon as they start following it, their health will start improving. It could be anything from gluten avoidance through to the breakfast–jelly bean–sandwich diet, the effects will be the same. Things would have improved anyway, but they will attribute the change to the plan. They will exalt their newfound diet from the highest rooftops, tell all their friends of the miraculous improvements "caused" by the plan, and leave glowing reviews on the blogger's website. They will believe, and, influenced by a growing number of positive testimonials, so will many others.

Of course, some people will discover the diet when they are not at the lowest point of their health journey. Perhaps because they are feeling OK when they start, or maybe, if they have a more serious condition that is not likely to improve, they will not be convinced. The likelihood is that they will just wander away from the plan. They will not leave testimonials (when have you ever seen an ambivalent testimonial?) and their views will not be taken into account. If you look on the blogger's website, all you will see is a long list of testimonials that tell of the miraculous power of the new diet. Even to the most cynical among us, that story can sound very convincing.

IS ALL DIET AND HEALTH ADVICE WRONG?

I am not saying that every anecdote of diet-related health improvement has its roots in regression to the mean. Some treatments *do* work. If changes to the diet did not have any effect on health, there would be no science of dietetics. But in many cases, reported health improvements may well be just those that would have occurred anyway. Without carefully designed experiments we cannot pick apart correlation and causation.

There are, however, clues that might help us identify regression to the mean in action. Did the initial improvement occur during the blogger's own period of ill health? Was it shown to work in controlled experiments, or is all the evidence being presented from anecdotes and testimonials?

Learn to spot the effects and you will start to see it everywhere. You will see it in miracle diets and the testimony of naturopathy patients. You will spot it in homeopathy, superstitions, and witchcraft. It can be seen in sports, fitness, education, and health care. It can explain the vast numbers of antibiotics demanded by patients with nonbacterial illnesses (even though antibiotics don't work on viruses).

Once you can see it in action, regression to the mean can set you free from that nagging doubt that there might just be something in all of this pseudoscience. It can explain why so many bright people seem convinced by utter nonsense. Those falling under the spell of this deceptively simple effect are not necessarily stupid. Just remember that it was not even categorized until the late 1800s. It is frequently missed by many of the world's smartest minds, and we are all capable of being deceived by its magic.

Thankfully, science can design experiments to negate regression effects and definitively spot whether correlation is in fact causation. The success of the scientific method and the reason for the staggering progress of humanity since its creation is that it manages to see past the pitfalls and searches for hard evidence. But for scientific progress to continue, we need to have faith in its method. Too many in the media, the public, and even scientists themselves are inclined to accept the quick answer.

If just a few more of us can get better at spotting regression to the mean in action, it might become a bit harder for false beliefs to take hold. At the extremes, a misunderstanding of this effect can have terrible consequences. False beliefs can do great harm in the real world. Sometimes, they can even cost people their lives. If a few of us making the effort learn a bit of statistics that will help to counter these beliefs, I think that is a price worth paying.

By the way, a quick note about the Voltaire quote at the beginning of this chapter: It is clearly from a different time, and these days, there are plenty of evidence-based treatments available to the world of medicine that provide more than just distraction. Perhaps we should change the phrase to "The art of *alternative* medicine consists of amusing the patient while nature cures the disease," which is something we shall discuss at length in later chapters.

Chapter Five

THE REMEMBERING SELF

ANGRY CHEF RUINS THE WHOLE EVENING, AGAIN

Throughout our lives we will experience the world in two ways, according to the behavioral economist Daniel Kahneman (I know I've mentioned him before, but he's really good). The experiencing self is a fast, intuitive, and largely unconscious mode of thinking that operates in the present moment. It monitors the world around us, deciding how much we like certain things, how much pain, fear, heat, taste, smell, pleasure, or boredom we are feeling at any particular moment. If someone asks you how hot it is, or if it is raining, your experiencing self will provide the answer. The remembering self, on the other hand, is the part of the brain that takes information from the experiencing self and turns it into the story of your life. It is slower and more considered, and defines how you think about the stuff that the experiencing self has done. It keeps the score and when someone asks you, "What has the weather been like this summer?" "Do you enjoy your job?" or "How was your day?" it is your remembering self that will give the answer.

Understanding these two ways of viewing the world can shed light on why some of us are so inclined to false beliefs about food, and how those beliefs can become so deeply held. It can also explain how each of us experience the same world in a completely different way, because although our memories may be very powerful, we do not have the storage space to remember every single thing.

The experiencing self sees the world in very short flashes, maybe only lasting about three seconds, and most of the time these moments disappear almost instantly. The only things that get transferred across to the remembering self are the important bits, the significant moments, the changes to the story. Because of the limited storage space, the remembering self is quite selective as to which little flashes of experience it will include, with many less significant moments vanishing without a trace. This means that the remembering self will tend to color our experiences with the impact of something significant. It will often ignore happy moments of quiet contentment, leaving them to disappear into the ether. The remembering self likes to focus on the extremes.

Just imagine we are having a nice evening out with friends and family in a much-loved local restaurant. Everything is going perfectly and everyone's experiencing self is having a lovely time. The service has been impeccable; the food, flawless; the drinks and conversation are flowing freely. Somehow, toward the end of the night and through no fault of my own, I accidentally manage to get really drunk, insult Mrs. Angry Chef's sister, fall over a chair, and spill a drink into her very expensive bag. It's the sort of thing that could happen to anyone, and I am sure we all agree that no one should attach any blame and there should be no recriminations should this completely theoretical example ever happen.

"I can't believe he did that; it ruined the whole evening," everyone will say. "The whole thing was a complete disaster," they will declare. Of course, it did not ruin the whole evening—everyone's experiencing self still had a nice time for the majority of it—but the remembering self has the tendency to cloud the whole thing because of the extreme bad memory at the end. Similarly, when asked to remember what the weather was like last summer, our memory may well be clouded by a couple of fairly extreme days. A short heat wave when the air conditioning was broken or a difficult rain-soaked camping trip will cloud our judgment and the remembering self will answer accordingly. When asked about the past, we create a story based on the limited bits that we remember, generally the significant points in the journey. Crucially, the remembering self will make a judgment as to which things are important, so while Mrs. Angry Chef is likely to think of the whole embarrassingly drunk husband thing as the most significant point in a story, a small child experiencing the same evening might be more likely to say that a particularly nice strawberry ice cream was the most important memory. The remembering self is the one who decides how the story is told, and this will depend greatly on our beliefs, values, and opinions. We all tell our own story and one person's vital plot point is another's throwaway memory. We pick our own extremes, and these define how our story is told.

It is our remembering self that guides our future decisions, so this clouding by extreme events does have an effect on our behavior. Our brains are wired to remember intensely negative experiences and adjust how we act to avoid them, even if this means missing out on subtler positive times. This has probably helped our survival as a species, but it is not particularly useful when it comes to maximizing our day-to-day happiness. This is what Kahneman calls the "tyranny of the remembering self" and it can result in us leading less contented lives. We may well be likely to miss out on many moments of contentment so as to avoid even the tiny possibility of something terrible.

THE POWER OF BELIEF

It has long been curious to me that intelligent, educated people can fall for pseudoscience and nonsense so completely. I have discussed some common food myths, such as detox and the alkaline diet, but to be successful, these beliefs need to be sustained over a long period of time. When your business involves selling certainty, you do need to be very certain, and in this modern information age, that must surely be a very hard thing to achieve.

Take the alkaline ash diet. As we have discussed, the science on which this is based is so flimsy as to be laughable, but to its many proponents it is based on stone-cold fact. When I look into the eyes of those who evangelize about it, I can tell they truly believe. It does not make the results of following the diet any less dangerous, but I do not believe they have any malicious intent.

Yet when the diet's founding father has been convicted of practicing medicine without a license, the advocates must be aware of the criticism. When your living depends on something, I can't believe you don't Google it occasionally. When it's this hard to find a dietitian, medical doctor, or respectable medical authority willing to support it; when the internet is awash with incredulous criticism; and when just a few moments of research would reveal a huge amount of informed debunking from all manner of respectable quarters, wouldn't most people waver in their certainty? In any search for evidence we will select for belief-confirming information based on preexisting biases, but this is a different level of self-delusion, and I am deeply curious to know how this complete, unwavering certainty can be maintained. Not only do you need to find flaws in the negative arguments, you almost need to act like they don't exist.

FORMER-NATUROPATH-TURNED-PSEUDOSCIENCE-DEBUNKER BRITT MARIE HERMES

After talking to Britt Marie Hermes, I started to think about how the remembering self might guide our beliefs. Britt's story is particularly interesting because it offers a rare glimpse into the world of false beliefs from someone who, by her own brave admission, once held them.

By any estimation, Britt is bright and informed. She has a master's in molecular biology and is now studying for a PhD, which will look at the reasons why people make health-care decisions. But just a few years ago, Britt was a practicing naturopath, dispensing ineffective treatments and quack remedies, completely free from any doubt and entirely unaware that she might be selling a lie.

Britt grew up in Southern California, an area of the world with a well-publicized obsession with health, body image, and new age belief systems. In her younger years, food and health were a huge preoccupation, driven by body-image issues and a need for control—hardly surprising when your formative years are spent in the image-fixated birthplace of almost every food fad on the planet. Britt talks about how layer after layer of positive anecdotes and experiences regarding alternative medical practices, including her own recovery from an autoimmune condition, made acceptance of these beliefs very easy. She spent some time working with a highly educated and otherwise informed family who, based on their own experience, believed in the causal link between vaccines and autism. This provided more reinforcement for a belief system that challenged the accepted view of "conventional medicine." Her upbringing and youthful experiences strongly inclined her toward alternative medicine, so much so that when she made the decision to train as a "naturopathic doctor," she firmly believed that she was entering a legitimate branch of the medical profession. Her parents questioned why she did not just go to medical school, but in Britt's mind she was simply entering a more dynamic branch of medicine, one that more accurately matched her existing values and beliefs. "I genuinely believed I could practice medicine and go into a large public health field or work for the WHO," she tells me. "That was always my dream, and I now realize I need to redo my whole education to achieve it."

Once she started studying, she began to accept more and more unscientific fallacies without question. She was in a classroom setting, learning from experienced authority figures. Naturopathy, which encompasses bizarre and pseudoscientific practices such as homeopathy and crystal healing, is not taught in a tepee by cross-legged gurus talking about their spirituality—if it were, perhaps fewer people would fall for it. In Britt's case, the established educational setting, the veteran and seemingly learned teachers—the trappings of educational respectability—were all signs that this was a place of truth.

We shall revisit Britt's story in chapter 21 (on fighting pseudoscience), and talk a little about the years she spent "treating" patients, but when I was speaking to her, there was one question that I really struggled to answer. How could someone educated, informed, connected, interested in science, and determined to practice medicine spend so long living with such ingrained false beliefs? Before this world came crashing down, Britt has absolutely no memory of there being any internal conflict. She cannot remember reading anything that countered her naturopathic beliefs. Even when questioned about her career choice by her parents, she simply explained it as her choosing a particular path within the field of medicine, surely something that a bright, career-minded person would have researched in detail. For the whole time she studied and practiced, she never had a moment's doubt.

I cannot imagine how this could have been. It seems incredible to me that someone could have existed for so long with such false beliefs and never doubted for a second. I sometimes look into the eyes of the latest diet gurus as they tell me that dairy leaches calcium from your bones, or lemon water is alkalizing for your body, and I try to look for seeds of self-doubt in their eyes. Surely now and again they come across sciencebasedmedicine.org or one of the many other resources giving out sensible advice. I am often left assuming that they must be incredible actors, because I never see even a flicker to give them away.

Maybe the reason for this is because, like Britt throughout her years in that world, they genuinely have no doubts. Surely that is the only way pseudoscience can flourish so effectively. Doubt, uncertainty, questioning—they are the qualities of science and its unceasing search for truth. For pseudoscience to flourish, perhaps we must believe in certainty?

Fooled by the remembering self?

Maybe some of the answer lies in the remembering self. With the plethora of information available to us in the modern age, I cannot imagine that Britt's experiencing self did not contain some moments, some conversations, some seeds of doubt. Perhaps a friend mentioned that the whole premise of homeopathic treatment is completely crazy. Maybe she read a newspaper article about the false claims of antivaccine campaigners. Surely the fact that some therapies and treatments are just so fucking stupid must have triggered something in a bright and active mind. It is my guess that for Britt, and many others on the same path, the experiencing self had some of these moments. But it is quite possible that our upbringing, lifestyle, belief systems, and many other factors condition the remembering self to select only the points in the story that it sees as relevant.

In my case, purely because my early life took a different path, I would have been conditioned by comments from doctors, scientists, skeptics, and naysayers. But for Britt, maybe comments like these were rejected, not to be included in the story of her life. The fragments of memory that could have saved her from years of wasted education were casually discarded by her remembering self in favor of the anecdotes that reinforced them.

Fortunately, as well as a susceptibility to false beliefs, Britt's life experiences also gave her a strong sense of moral responsibility. More than anything, she wanted to help and care for patients and make a difference in the world, and despite taking a path in alternative medicine, she maintained a respect for authority. When packages containing a drug called Ukrain went missing in transit, her boss mentioned that they had probably been confiscated. For the first time in her career alarm bells started to ring, and when she started to research Ukrain she found that it was not approved by the Federal Drug Administration, making it a federal crime to import it into the United States. Shocked, Britt confronted her boss, who had been using the drug in the treatment of cancer patients, many of whom were terminally ill. He admitted that his actions were "legally questionable," leading Britt to resign and hire herself a lawyer. Immediately, senior figures from the naturopath community started to contact her, urging her not to go to the authorities.

After former colleagues and friends turned their backs, Britt started to look in more detail at the training she had received as a naturopath, eventually reaching out to a number of people within the skeptic community (including Professor Edzard Ernst and Dr. Simon Singh, authors of the excellent *Trick or Treatment?*, a book that investigates alternative medicine). It did not take long for her to realize that naturopathy, something she had dedicated nearly eight years of her life to, was far from the dynamic branch of the medical community that she believed it to be.

Once this all came crashing down, the way she perceived the world was changed forever, which must have been a traumatic and life-changing experience. But incredibly, this did not alter the way she views the past. Even a complete resetting of her belief system did not open up the lost and discarded memories, and the story of her naturopathic years is one of complete, unwavering belief.

I see this in the false belief systems adopted by many of the self-appointed health gurus we have discussed. Many are taught, in austere educational settings, the benefits of detox and alkaline eating. They tend to have an existing background in new age thinking and connections to the world of naturopathy. Almost all are hugely affected by powerful personal anecdotes. Their lives have primed them for acceptance, but to make that acceptance sustainable I believe that their remembering selves carefully select moments and experiences that confirm their existing belief systems, discarding negative comments as trivial and not of concern.

The beliefs we hold about food, diet, and health are likely to be guided by our remembering self, which can be very much the tyrant that Kahneman describes. The only trick available to us is to be very aware of which memories we keep and which ones we discard, because if we discard the wrong ones, it might just leave us with a highly distorted view of the world.

Part II

WHEN SCIENCE

GOES WRONG

Chapter Six

THE GENIUS OF SCIENCE COLUMBO

*O*ne day, we will all surely die.

Oh good. Another gentle, heart-warming chapter then.
The most we can hope for is a little peace and dignity in our final
hours. When my time comes, I am hoping that death's icy grip
will not take hold after days of unremitting stomach pain, vomiting, fishy-smelling diarrhea, hallucinations, and seizures. That
is exactly the fate that befell fifteen thousand Londoners in 1849
in one of the worst cholera outbreaks of the time. Cholera was a
vile blight on the new metropolis that London had become and,
despite the huge industrial and technological advances of the age,
there was very little available in the way of treatment. To make
matters worse, no one really had much of a clue as to its cause or
how such devastating outbreaks might be prevented in future. The
human and economic toll of this disease was huge, and as these
hugely destructive incidents increased in frequency, they risked
stopping the urbanization of London in its tracks. Luckily, science
was there to save the day! But before I explain exactly how that
happened, it's probably worth introducing a new character.

INTRODUCING SCIENCE COLUMBO

Science is frustrating at times. Just when we have a great story, something we can all hang on to, science comes along with its pesky, unstoppable quest for facts and evidence. Sometimes, it reminds me of Peter Falk's TV detective Columbo. When everything seems to be in place, with all the pieces fitting nicely together, Columbo is always bothered by "just one more thing." The same is true of science. Even when we are talking about the most fundamental concepts imaginable, Science Columbo is always troubled by "just one more thing."

Take Newton's law of universal gravitation. Things don't get much more fundamental than that. Published in 1687, it quickly became accepted as a profound work of genius. It is engagingly simple, stating that any two particles in the universe are attracted together with a force dependent on their masses and the distance between them. The bigger the two masses and the closer they are together, the bigger the force.

Most of us can understand this fairly easily. It helpfully explains why we don't fly off into space, the gravity caused by the earth's huge mass and close proximity securely holding us in place. It can be used to predict how fast an apple will fall to the ground, the trajectory of space flights, the orbits of planets, and lots of things that we observe in the universe. When combined with an engaging story of discovery set in an idyllic Kent orchard, we have one of the most compelling and powerful science stories of all time. With a moment of inspiration, a lone genius explains a fundamental force that shapes everything.

Then along comes Science Columbo.

Science Columbo: So, Mr. Newton. Looks like this law of universal gravitation has worked out pretty nicely for you. Everything seems to add up. It all makes sense. The earth's orbit around the sun. The moon's orbit around the earth. The reason we don't all fly off into space. You must be pretty happy with it. I best be on my way. Good luck with everything.

Newton: Thanks, Science Columbo. See you around.

Science Columbo is on his way out of the door. Halfway through he stops and turns back to Newton, raising a finger.

Science Columbo: There's just one thing that's bothering me. I wonder if you can help me explain it?

Newton rolls his eyes.

Newton: What's that, Science Columbo?

Science Columbo: I can't get this out of my mind. Aren't there these tiny variables in the orbits of Mercury? If your law really is universal, then it doesn't really fit. Mercury's orbit couldn't possibly change if its mass didn't change. Can you help me explain that Mr. Newton?

Newton: Experimental error, I guess. I mean, come on, it's only forty-three arc seconds per century. It could easily just be a mistake in the measurements.

Science Columbo: Maybe. You're probably right. But I just can't get it off my mind.

And there we have it. Science Columbo sees the slightest problem with the theory and just can't stop picking away. Tiny variations in the orbits of planets close to the sun did not fit with Newton's theory (although, to be clear, this was not discovered until long after Newton's death, and Science Columbo is not real). At the time of discovery, many people thought these variations small enough to be explained away, but much like the real Columbo (by which I mean the fictional TV detective), Science Columbo would not rest until everything was resolved. Despite huge resistance from the scientific establishment, who were keen to maintain the universality of Newton's law, Science Columbo would not let go. Eventually this led to Einstein developing his theory of general relativity, which countered Newton and explained the tiny discrepancies in the orbits. Not long afterward Arthur Eddington, very much the unheralded hero of the piece, designed a fiendishly clever experiment to test Einstein's theory, using a total eclipse to measure the way the gravity from the sun bent light from distant stars. And Science Columbo, with a little help from Einstein and Eddington, was proved to be correct all along.

Back to London: What occurred after the 1849 cholera outbreak is perhaps one of the greatest examples of Science Columbo's power for good. His actions, or to be more accurate, the actions of two men working under his influence, led to the combating and near-elimination of cholera epidemics in the developed world.

ARGUING YOUR WAY TO THE TRUTH

William Farr is regarded as one of the founders of medical statistics and one of the great mathematical minds of his age. In 1849, he thoroughly believed the scientific consensus on cholera—that it was transmitted through the foul London air. The city at this time was pervaded by the vilest stench imaginable, especially during the summertime, with desperately poor sanitation and a huge buildup of human and animal waste creating an unavoidable odor, said at its worst to be so acrid it burned the eyes. It was perhaps inevitable that a narrative had been constructed linking the stench—the most unpleasant part of day-to-day city life—with the disease, the vilest and most spiteful part of urban existence at the time. Both were strongly associated with dirt, shit, and squalor. Both occurred together. The smell was a huge, obvious, unavoidable hare that sat next to the disease, a pile of shitty, pestilent eggs.

It is worth remembering that at this time there was still no germ theory of disease, so the majority of Farr's work involved the collecting and interpretation of data. He painstakingly compiled information about occurrence, the prevalence in different areas, how it spread, and how it affected population groups. He was meticulous and brilliant, in many ways creating the field of medical statistics, something that would go on to change our understanding of disease and save many millions of lives.

Farr analyzed unprecedented amounts of data and hundreds of potential variables. He produced a paper after the 1849 cholera outbreak that showed a strong correlation between the elevation above sea level and the likelihood of contracting cholera. This seemingly confirmed the prevalent "foul air" theory of transmission. It was a beautiful and groundbreaking piece of research that, in any other field, would have been the definitive proof people had been looking for. However, this was the world of science and in science, there is always . . .

Science Columbo: There's just one thing that's bothering me Mr. Farr.

William Farr: What's that, Science Columbo?

Science Columbo: There's this other guy, goes by the name of John Snow. He has some data, too, but it shows a link to water sources, not elevation. It doesn't seem to make any sense to me.

William Farr: But I have more data. And Snow is just an outlier. No one really believes him.

Science Columbo: Maybe, but I just can't get it out of my mind.

John Snow, another statistician, did not accept the foul air theory and had demonstrated some strong statistical links to water sources for the 1849 outbreak, postulating that some sort of pathogen ingested in contaminated water was breeding in people's guts and causing the disease. This was largely rejected at the time, with most preferring the more-compelling foul air theory, but because of the way science works, it was not completely discounted. William Farr, considered by most to be the leading authority, was wise enough to accept at least the potential for his own ignorance. After a further outbreak in 1853, Snow produced more compelling evidence that water from two particular sources showed strong correlations to the disease, and Farr, along with the rest of the scientific community of the time, began to accept that contaminated water may well be one of the potential causes.

By 1866, sometime after Snow had died and after numerous confirmatory studies, Farr accepted that the water supply, not the stench, was the way cholera was transmitted. This work led to the creation of sewer works and water processing, and in little more than fifty years, the scourge of cholera outbreaks had been virtually eliminated in the developed world. This understanding saved countless lives and enabled the safe urbanization of populations throughout the world. Although cholera is still a blight on humanity, costing many thousands of lives each year, at least now we understand its cause. Where outbreaks occur, if the resources exist, action can be taken to mitigate disaster.

The battle between the competing theories of Snow and Farr should not be seen as a battle to see who had the greatest scientific mind. Snow's theory had the advantage of being correct, and it is he, perhaps rightly, who is thought of as the man who saved so many millions of lives. But we should also give credit to Farr, thought by many to be the greater statistician. It was Farr who proved himself capable of one of the hardest of intellectual disciplines—the ability to accept that he might just be wrong.

Here lies the genius of science. If the prevailing theory of the day had been accepted after Farr's research into the 1849 outbreak, many resources might have been spent clearing shit from the streets, perhaps forcing even more of it toward the water supply and costing many more lives. But it was not, because there was a willingness within the scientific community to accept the potential for their own ignorance and keep looking for evidence. Eventually, with enough evidence and open minds, the truth was found. This is how science has changed the world, how it has cured disease, advanced technology, and increased our understanding of the universe in ways that could not have been imagined in Snow and Farr's time. That is why I love science as much as I love food, perhaps even more.

THE PROBLEM WITH SIMPLE STORIES

But there is a problem. For a moment, let's imagine how the modern health blogging generation might have reported the period between 1849 and 1866, when vicious cholera epidemics were costing thousands of lives. In a Victorian Goop special feature, they might have said:

Why can't these so-called experts just stop all this infighting and make up their so-called minds? It is obvious that this so-called cholera is caused by the smell in the poor stinky parts of London. You only have to go there to know that. It smells really bad and you can just tell that it is making you ill when you breathe in that toxic air. I have so-called friends who live in parts of the country that don't smell nearly as bad and none of them have cholera, so it must be the smell causing it. It is about time that they started moving all that poop out of the way. If they just drop it into the Thames, it will go into the sea, and the stench will be gone forever. They need to stop arguing about their so-called statistics and get busy with some so-called shovels.

If it had been suggested that processing water might be the best way to prevent outbreaks, a further comment piece might have read:

OMG, I cannot believe that these so-called experts are now saying we should be drinking "processed" water. Everyone knows that processed things are just so bad for you and full of evil toxic chemicals. I for one will not be found drinking anything processed and will stick to the same natural unprocessed water that my grandmother used to drink. She lived to be thirty-five and never once had cholera.

I realize that by creating a past where bloggers are also in charge of science, I have granted them some sort of immortal megapowers, but hopefully you get my point. Science is not broken when people disagree. In fact, disagreement is a fairly good sign that it is working. Unfortunately, this need for uncertainty to drive scientific progress is hard for many of us to grasp.

So, who can we blame?

I guess we should blame our education system. I don't think I have blamed that for anything yet, so it's probably about time. Science education generally revolves around the teaching of scientific facts. It very rarely touches upon the more profound and interesting aspects of science that might actually be able to help people in their lives. After years of education, most of us are left knowing little about the scientific method, the need to accept uncertainty, the need to look for and respect evidence, and an understanding of what constitutes proof.

Paul Rozin, a professor of psychology at the University of Pennsylvania whose work focuses on food behaviors, has studied extensively the reasons why nonceliacs adopt gluten-free diets, but his work has shown that even undergraduates are not educated in a way that helps them understand evidence. He has also shown that most are comfortable with the word *proof* being used very loosely. Many of the people he has studied remain completely unmoved when presented with research showing that gluten-free diets will not benefit them, preferring instead to believe salespeople, advocates, and advertising:

> Most of the people following the diets are evidence insensitive. Many following gluten-free [diets] don't even know what gluten is. This makes it easy for small things to be blown out of all proportion. It was the same with the oat bran fad. There was a little bit of science behind it, some evidence of a small benefit, but people want to think in terms of big benefits. People are not trained and educated to respect evidence.

Science in schools is taught as a series of black-and-white facts, looking at things that are definitely true and trying to explain the reasons why. Cholera is caused by a bacterial infection of the small intestine. Energy and mass are always conserved. Yawns are contagious. Sugar is a carbohydrate. The earth orbits the sun. Antioxidants attack free radicals. That sort of thing. This is a valuable pursuit, and I am a huge fan of facts, but it does come with a number of problems. We are led to believe that science knows all, that it possesses a large set of definitive truths and its goal now is simply to share and explain those truths to the public.

Science does indeed possess a lot of very interesting facts. But at the edges, at the coal face of science, there is always going to be uncertainty and doubt. The interesting parts of science are where the disagreements are, and when there are disagreements, the public is likely to be left confused. We are easily led by a disconsolate media to believe that science is broken. This doubt and ambiguity is likely to leave the instinctive brain unsettled, because if there is one thing it hates, it is uncertainty.

That's true. I wasn't sure if we had any cookies left the other day. Had to go and check.

This uncertainty will leave it searching for other, more certain messages, even if that means trusting an unreliable source.

THE PROBLEM WITH FOOD

A big problem for anyone studying nutritional science is that designing experiments to accurately test the effects of diet on health is really, really hard. A lot of experiments are too basic to tell us much, if they are testing single nutrients on cells or tissues in the laboratory or looking at effects on animals. Unfortunately, the human body and the composition of the food we eat are both staggeringly complex, and often, reality refuses to play along.

If you want to conduct nutrition experiments on human volunteers, there are a number of different problems. First, you will need to control their entire diet. You could potentially keep them captive in a laboratory for a period of time, carefully controlling and monitoring what they eat, but obviously there is a limit on how long you can keep people contained. If you are looking at specific changes over short periods of time, then you might be able to get some fairly good data, but nutritional science is more interested in long-term effects. You cannot just take short-term changes and predict long-term outcomes.

Let's say you are studying the effects of beets, and you want to conduct a placebo-controlled study. It is not going to be possible— because there is never going to be a placebo that looks and tastes like a beet but has zero effect on the body. Anyone in the beets group will know what they are eating, so they will be susceptible to any placebo effects that might occur. If you want to look at long-term changes, you might try to make some sort of intervention, such as giving people a supplement to add to their diet, asking them to cook with a particular oil, or supplying them with a special ingredient every week for a few years, but this is likely to be expensive and could produce tricky confounding factors. For instance, you would need to account for what the new ingredient was replacing in people's diet and think about whether people might make other changes to their lifestyle when they know they are being monitored. It would also almost certainly rely on self-reporting of food consumption. As you cannot study people the whole time, you need to be sure they are not just throwing away the new ingredient or feeding the supplements to the dog.

The other way to study such things is to conduct detailed epidemiological studies on diet and health of the sort pioneered by Snow and Farr in the 1850s. You could look at statistics on people's food consumption and try to find correlations between different diets and health outcomes. As we saw in the case of the cholera studies, this is fraught with problems and inaccuracies, too, with much care needed to look for hares and find the lapwings. This sort of study is increasingly sophisticated, and technology has made it many times easier to sort large data sets than it was 150 years ago, but there are still problems. Although many health outcomes can be measured empirically and medical records will be available, it is fairly unavoidable that this sort of study will rely on self-reported data when it comes to knowing what people eat. Unfortunately, this is notoriously unreliable. When reporting what we eat, we are all inclined to be a little creative, especially when we know nutritional scientists will be studying the results. Anyone who has ever been asked by their doctor how many units of alcohol they drink every week will appreciate that this might be the case. ("I know I had forty units this week, but I had a detox smoothie and a couple of green juices on Sunday. Let's say fifteen.")

For these and many other reasons, studying the effects of different nutrients on health is extremely hard, and studying the effects of different diets is even harder. There will always be a fair degree of randomness built into any data, and with randomness comes the possibility of unusual results. Statistician and professor David Spiegelhalter says, "I take little notice of single studies—they are often not very good science and get publicized because they happen to come up with a surprising result. And surprising results are usually wrong."

And herein lies a huge problem. Interesting results do pop up now and again in nutritional science, but often these occur because of the inherent inaccuracies. The more interesting a result is, the more likely it is to be wrong. To make things worse, interesting results are often the ones that reach the public, as they are the most likely to be reported by the media. And when the media reports them, they are sold to the public in much the same way that school science is, as stone-cold facts.

Here is a selection of newspaper headlines taken from mainstream news sources over the past few years, collected by the excellent website Kill or Cure, which gathers news that attempts "to classify every inanimate object into those that cause cancer and those that prevent it":

- Cheers! Now They Tell Us Beer and Wine Give Us Cancer
- Why Beer Is the Latest Hope in the Fight Against Cancer
- High Fiber Foods to Help Fight Cancer
- Fiber Cancer Risk Warning
- Even One Drink a Day Can Up Breast Cancer Risk
- Drinking Two Glasses of Wine a Day Can INCREASE Risk of Cancer by 168%
- Wine "Helps Prevent Cancer"

All these stories report the conclusions of real scientific papers, produced by completely legitimate scientists from around the world. They are all interesting results, worthy of a story in a national newspaper, but clearly they cannot all be true. There is a fair amount of misreporting going on, which I shall get to later, but the main flaw lies in the fact that the newspapers are understandably inclined to focus on the interesting bits of science, the outliers, the bits where the arguments and disagreements are likely to happen. The bits most likely to be wrong.

Ah. So, it's the newspapers' fault then. Knew it.

Maybe. A bit. No one ever minds if you blame journalists. But it is hardly surprising that these sort of stories are reported so frequently. The quality of nutritional studies can vary a great deal, and it is pretty much impossible for laypeople or journalists to be able to accurately distinguish between good and bad studies. All the information reported comes originally from the scientists who created it, and these are people who should really know better.

Perhaps the way science is funded and assessed doesn't always help. The work of British academic institutions is assessed through something called the Research Excellence Framework, and around 10 to 15 percent of the assessment depends on something known as the impact of the previous year's research. This is a self-reported and largely subjective measure, but much of it is focused on the media impact of that year's research. Being featured in international news media, or being featured on a nationally aired television program, is a great way to boost your impact score and your chance of future funding. Universities, all of which will have eager press departments, will be keen to take the latest research and issue press releases to frame it in a way that will be likely to get media coverage.

Also, it is extremely likely that some scientists out there enjoy the media spotlight. They want to frame their life's work and passion in a way that oversells its importance to the world. In food, this is a particular problem. Right now, obesity is a huge and poorly understood health issue and that means, for people studying it, that the stakes are extremely high. Arriving at a solution or a cause could easily result in worldwide acclaim. In addition, unlike in some other areas, many different fields of science have an interest in obesity. Behavioral economists, psychologists, geneticists, evolutionary geneticists, paleontologists, neurologists, microbiologists, biochemists, molecular biologists, epidemiologists, genetic epidemiologists, dietitians, and many others study obesity, and all will have a different approach. Scientists, and particularly highly specialized scientists, are not beyond a susceptibility to bias and may be inclined to believe in narratives that fit their view of the world. It can make for a confusing picture, where seemingly contradictory messages fly around all the time.

Ah, so it's the evil scientists all along. Confusing us with their facts and research.

Sometimes. But reporting of science can be irresponsible, too. A number of scientists writing for newspapers and magazines are capable of composing highly persuasive articles, and others with a television profile are capable of making a strong and vociferous case for their own field of study. Often this can skew the public's mind to believe that a particular area of science holds the answer.

DON'T BELIEVE THE HYPE

The drive to create a simple, compelling narrative from a set of facts can also lead journalists and headline writers to bend the truth. Dr. Javier Gonzalez researches nutrition and metabolism at the University of Bath. In 2016 he published research that showed ingesting sugar solution reduced glycogen depletion in the liver during intense exercise. A lay summary of the work was approved by the research team and released by the university press department, but when press reports appeared, they all followed the line that scientists had proven that sports drinks are a waste of money and you might as well just drink sugar water to improve sports performance. In the following days, Javier was asked in a television interview: "Has the sports drink industry wasted our money in the last few years?"

Javier told me:

Clearly this was not the case. The study did not compare sports drinks to sugar. We also didn't measure performance; we were just looking at mechanisms. The media do have a tendency to extrapolate and hypothesize too much, looking for the headline. It was nice to get the exposure, but it is disappointing to see your research misrepresented. You worry about reputation, but to be honest, other academics understand.

A number of other academics I spoke to are frustrated by the role that television plays in misrepresenting nutritional study. Most nutrition researchers are frequently approached by television companies not to talk about their research but to help conduct in-program experiments on small groups. Javier also told me, "I saw a program recently and they said 'science has shown that X affects Y, but let's see if it works for real.' As if science was somehow not looking at the real world. They then went on to conduct an experiment on five people to show protein effect on muscle mass and presented it as evidence."

The tendency of television-program makers to conduct seemingly legitimate (but deeply flawed) experiments, often teaming up with publicity-hungry researchers, does come with a danger of reinforcing the public's misunderstanding of how science works.

So, it's the journalists' fault then.

Possibly. We all crave certainty, and we all want science to be black and white. Journalists fall into this trap, as do the public, but scientists are often equally keen to present themselves as having all the answers. Many scientists strongly identify themselves as the "holder of the truth," the fountain of knowledge, and the possessor of superior intellect. If only we lived in a world that valued the more profound sign of wisdom: the ability to accept the potential for your own ignorance.

So, it's the instinctive brain's fault. Again.

A little. Unfortunately, it is the way we are all conditioned to be. Since Snow and Farr's time, the public is considerably more engaged and connected to the world of science. Unfortunately, our increased education has not always led to a greater understanding of the way science works. This causes problems, and despite its huge and unremitting progress, we are often led to believe that science is broken. We must all learn to accept a little more uncertainty and doubt and understand that the fallibility of science is a strength and not a weakness.

Good. That's nice. I will try to remember. Accept uncertainty and doubt. Science is usually wrong.

. . .

Hang on a minute.

What?

Aren't you always saying that we should trust dietitians, medical doctors, and registered nutritionists? And they are all scientists. Does that mean they are all wrong? I hope so, because that dietitian we know told us off for eating too many cookies.

Ah, I'm glad you brought that up. Clearly, when it comes to living our day-to-day life, we need to make some decisions. We can't just eat everything, blaming science for not making up its mind. Unfortunately, in this chapter I seem to have explained why science is full of indecision and doubt and we should not really accept individual pieces of research as being a definite new truth. This does beg the question "What is the point of science, and how can it help us live our lives?"

Yay! Science is wrong. Pass the linzer tortes.

Public health guidelines do exist. Such bodies as the National Institutes of Health (NIH) and Centers for Disease Control (CDC) have a fairly consistent line as to what we should eat to be healthy. Incredibly similar guidelines are issued by numerous charities, governments, and public health bodies around the world, including the World Health Organization.

To a curious public, it is perhaps something of a mystery how such consistency can be achieved, despite the existence of numerous campaigners citing evidence that counters it. When science is by its very nature uncertain, how can so many people be so certain as to what we should be doing? David Spiegelhalter, someone who understands a bit about risk and the public, says that "when deciding on public health policies, it is essential to look at the totality of the evidence using what is known as a systematic review." Public health bodies issue their guidelines by assessing all the available evidence and combining it together to get an idea of the current state of scientific knowledge. With care taken to avoid the potential for bias, the importance and relevance of all serious studies into diet and nutrition will be assessed, and models will be produced by highly qualified and intelligent people using processes way beyond most of us. As I mentioned, unless you are very specially trained, it is difficult to assess the quality of scientific evidence, and when different studies have directly contradicting results, it can be even more confusing. Combining everything together in a systematic review is the best way we have of assessing the current state of scientific knowledge, and so it is the best hope we have for making decisions as to how we should live. And that means not too many cookies.

Boo.

This all comes with many associated problems.

Too right. Less cookies. I mean fewer cookies.

Within the studies combined to create systematic reviews there will always be those that have messages countering the official advice. These will get discounted as outliers in the systematic review, but it is very easy for people to get hold of single bits of information and believe that this is proof that public health guidelines are wrong. This is frequently seen from advocates of the sort of high-fat, low-carb diets that we will see in chapter 8, often backing up their one piece of evidence with compelling anecdotes.

In fact, this reminds me of an article I recently read by the food writer Joanna Blythman about public health nutrition guidelines, complaining that science was broken. Apparently nutritional science has become riven with infighting, and the public has been left not knowing what to believe. It was certainly not a shock to read this, as similar things are reported all the time. However, she was not discussing the sort of noble scientific infighting that led to discoveries about cholera. It concerned a number of people challenging the accepted public health guidelines on nutrition by citing single studies and biased, poorly constructed reviews.

This provides public health bodies with a big problem. Systematic reviews are time-consuming and largely impenetrable to most of us. How these reviews are conducted is not easy to understand, and although they are the best evidence we have, they just do not appeal to our instinctive brains. In a debate on the dietary guidelines, Dr. Aseem Malhotra, a vociferous high-fat-diet campaigner, talked about a single Spanish study on dietary health and was quoted as saying:

This robust study provides yet more evidence to abandon the "low-fat" and calorie-counting mantra and instead concentrate on eating healthy and nutritious high fat foods. A high fat Mediterranean diet—which I follow and tell my patients to—not only doesn't lead to weight gain but is also the most protective dietary pattern against heart disease, cancer and dementia.

As a counter-argument, Dr. Alison Tedstone, chief nutritionist at Public Health England, was quoted as saying, "Our independent experts review all the available evidence—often thousands of scientific papers—run full-scale consultations and go to great lengths to ensure no bias."

Boring. I prefer the other guy. The one who says we can eat as much fat as we want. I like fat.

This is indeed the problem. The work that goes into public health guidelines is incredibly dull and poorly understood, and lacking in glamour, excitement, and media attention. While the debate is played out in public, we are likely to be instinctively drawn to stories, anecdotes, and interesting studies. And as we have discussed, interesting studies are highly likely to be wrong.

Debate in science is important. In fact, it is vital for progress. It is also interesting and worthy of being reported in newspapers. The problem is how it is framed to the public. Single experiments should not be presented as new information by which we should live our lives. Somehow we need to respect the knowing uncertainty that drives science forward and still believe in the unglamorous messages of public health practitioners and medical professionals. This will never be easy, but if we can spend more time in schools teaching the method of science rather than presenting a list of facts, that might help a little.

This brings me to **Rule Number 5 in the Angry Chef's Guide to Spotting Bullshit in the World of Food**: *They will tell you that the truth is not up for debate. They are never bothered by "one more thing."* Although there is beauty and power in learning more about the world, perhaps the true beauty of science lies in how it works, in the way it accepts the possibility of its own ignorance and constantly hunts for the truth.

Science Columbo: So, Einstein. It looks like this general relativity thing has worked out pretty well for you. At last we now have a theory that explains everything. I guess I'll be on my way.

Einstein: OK, Science Columbo. See you around.

Science Columbo starts to leave. He stops just before he reaches the door, turns, and lifts a finger.

Science Columbo: There's just this one little thing that's bothering me, Professor Einstein.

Einstein: Yes?

Science Columbo: I just can't quite work out how this "general relativity" can be reconciled with the laws of quantum physics. However hard I think, I can't work out how it might produce a consistent theory of quantum gravity. Can you help me with that?

Chapter Seven

COCONUT OIL

So far we have looked at a couple of examples of textbook nutri-nonsense, where scientific principles are completely discounted in favor of something profoundly mistaken and wrong. There are many more of these in the world, so apologies if I have not included your personal favorite. And on the web, the Angry Chef has a blog! I always welcome reports of new examples and look to debunk any significant ones as thoroughly as I can. I am using the examples in this book to give an idea of how the world of fad foods works, and to provide some clues as to how to spot bullshit when you see it.

In the rest of this section, we look at how bits of real science are misunderstood or miscommunicated in the process of selling us stuff. That stuff may be food products, fad diets, books, or even just website hits, but in understanding nutri-babble it is important to remember there is always something for sale, even if it is not immediately clear what that thing might be.

Confusion in the world of food is hardly surprising. I have a degree in biochemistry and twenty years' experience working with food, and I spend much of my time researching food and health stories, but if someone asks me which one is the good sort of cholesterol, I still have to look it up. If I struggle a bit, I imagine most people are similar, or perhaps a bit worse. After all, human nutrition is a huge and complex science involving many different disciplines, and there is no one alive who understands it all. It is no surprise that the public are often confused and that the media sometimes gets stuff wrong. As we have seen already, in places where there is confusion and misunderstanding, people are also susceptible to being exploited. When the reflective brain cannot easily access the answers, the instinctive brain is likely to be the one who takes charge.

David Spiegelhalter, the Winton Professor of the Public Understanding of Risk in the Statistical Laboratory at the University of Cambridge, told me, "When it comes to food, there are many different people trying to make us believe that different things are of harm or of benefit, and the public's decisions tend to be value-based and not evidence-based. People will always tend to follow those they trust based on values."

For these next few chapters I want to look at how science gets mangled, distorted, and presented in strange and misleading ways, forcing value-based judgments. A lot of our ability to be fooled depends on whom we trust and which values we hold, and as there are so many different opinions about food, sometimes we may place our trust in the wrong places.

Get ready. I write as the Angry Chef for a reason, and when science gets mistreated in this way, it really upsets me.

THE MOST UNLIKELY SUPERFOOD OF ALL

In my observance of food and health movements over the last few years, one of the most curious developments has been the inexorable rise of a particular super ingredient: coconut oil. It is the fat of choice for almost every single new age health guru, lauded for its remarkable culinary versatility and seemingly miraculous health-giving properties. Perhaps kale is the only other ingredient that has been hailed so vociferously. The popularity of coconut oil is hardly surprising when you look at the claims being made for this once-vilified fat. Here are a few typical examples:

From Deliciously Ella: Coconut oil isn't only loved for it's [sic] healthy fats; it also has lots of other amazing properties. It's very antibacterial, thanks to its lauric acid content. So when our bodies break down the fatty acids in coconut oil they can kill harmful bacteria, boosting our immunity. It can also be used for treating fungal infections and has proven to help people who suffer from candida.

From the impossibly shiny Kimberly Snyder, perhaps Angry Chef's all-time favorite US blogger, from her website kimberlysnyder.com: Yes, coconut oil is considered a saturated fat, but not all saturated fat is created equal! Unlike some other saturated fats, coconut oil is comprised of about 65% medium chain triglycerides (MCT), which is a fat that is rapidly absorbed into the bloodstream and burned as fuel for our bodies.

From the self-styled "rock star of the superfoods and longevity universe," David "Avocado" Wolfe, a helpful list of coconut oil's almost miraculous properties:

- **Anti-bacterial**—stops harmful bacteria and infections dead in their tracks
- **Anti-carcinogenic**—boosts immunity and keeps cancer cells from spreading

- **Anti-fungal**—destroys fungus and yeast that lead to infections
- **Anti-inflammatory**—repairs tissue while suppressing inflammation
- **Anti-microbial**—fights infection and inactivates harmful microbes
- **Anti-oxidant**—protects against damage from free radicals
- **Anti-retroviral/parasitic/protozoa/viral**—rids the body of lice, and parasites such as tapeworms; kills infection in the gut, destroys viruses that cause influenza, herpes, measles, hepatitis and more

Unsurprisingly, with so much going for it, coconut oil has also found itself with a raft of high-profile celebrity endorsements, giving it the sort of PR that money just can't buy. Madonna and Miranda Kerr say that they eat spoonfuls of the stuff as a health supplement, Jennifer Aniston uses it for weight loss, and Angelina Jolie has it as part of her breakfast routine. Gwyneth Paltrow claims she uses it as a mouthwash and sexual lubricant (separately, I hope).

What makes this even more surprising is that coconut oil was once considered a great dietary demon, a symbol of the evils of the food manufacturing industry. Referred to as "tropical oils," coconut and the chemically similar palm oil were used extensively as a cheap, versatile fat and developed a reputation as one of the most reviled of food ingredients. The fat in coconut oil is around 90 percent saturated, something that has been strongly linked to an increased risk of cardiovascular health problems.[1] The presence of large amounts of tropical oils, particularly in movie popcorn, where it was used in huge quantities, became a watchword for nutritionally poor manufactured products, the evil food-manufacturing industry deliberately filling us with harmful saturated heart poisons.

These days, in a dramatic rebirth and reinvention, it is one of the most remarkable superfoods on the planet. In *The Coconut Oil Miracle* naturopath Bruce Fife provides an extensive exploration of its incredible power. As well as describing proven abilities to promote weight loss, he also outlines its role in fighting disease, supporting his case with numerous powerful anecdotes. Coconut oil, it seems, has incredible antimicrobial properties and can be used to fight tooth decay, peptic ulcers, cancer, epilepsy, influenza, Alzheimer's disease, genital herpes, hepatitis C, and even AIDS. Bruce helpfully explains that the fatty acids contained in coconut oil make it deadly to disease-causing pathogens, yet harmless to human cells, making it a potential solution to the growing problem of antibiotic resistance.

OK, so you didn't really think it could cure cancer and AIDS, did you? But all this has to have come from somewhere. Surely people are not just making up these superfood claims from scratch. To understand exactly what is going on, it is time to look at a little bit of the science.

THE CURIOUS CHEMISTRY OF COCONUTS

What was all that about lauric acid and medium-chain fatty acids? To look at the claims being made about coconut oil properly, we first need to go through some basic fat chemistry. There are different types of fats, but culinary oils are mainly triglycerides. Triglycerides are made of three ("tri") fatty acids attached to a single glycerol molecule. Each fatty acid has a reactive acid bit on one end and a long tail of carbon.

Although most culinary fats are made of triglycerides and it is important to have some fat in your diet, not all triglycerides are created equal. The main differences are in the composition of the fatty acids, which can have different lengths of carbon chains and different levels of "saturation." Each carbon atom has the ability to form four chemical bonds. In fatty acids, each carbon is bonded to another carbon on each side in a chain, leaving two spare chemical "arms" to be filled (yes, thank you, I know I am oversimplifying). For the most part these arms are filled with hydrogens. When fats have two hydrogens attached to every carbon in the chain, with three hydrogens for the carbon at the end, there are no arms left to be filled and they are "saturated." Sometimes, when a couple of hydrogens are missing, a carbon can form a double bond with its neighboring carbon in the chain, making an "unsaturated" fatty acid. (If there is one double bond, you have a monounsaturated fatty acid, and if there is more than one, it is polyunsaturated.)

SATURATED FATTY ACID

O
H H H H H H H H H H H H H H H
C-C-C-C-C-C-C-C-C-C-C-C-C-C-C-H
H-O
H H H H H H H H H H H H H H H

UNSATURATED FATTY ACID

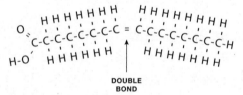

DOUBLE
BOND

Generally speaking, saturated fats, with all their carbons satisfyingly filled with the requisite number of hydrogens, are a little more chemically stable and harder for our bodies to metabolize quickly, so it is thought that these are more likely to stick around in the body with potential to do harm. Although I am highly aware of the irony of oversimplifying health and nutrition in a book where I expose people for oversimplifying health and nutrition, there is good evidence from systematic reviews* that high levels of saturated fat in people's diets increase their risk of heart disease. In general, sensible dietary advice around the world focuses on reducing the level of saturated fat in the diet and switching to unsaturated fats.

Simple. Or maybe not. Also important seems to be the length of the fatty acid carbon chains. There are two common types of fatty acids in most culinary fats: long-chain fatty acids (LCFAs) and medium-chain fatty acids (MCFAs). (Short-chain fatty acids are also present, but usually they are at low levels and not as important in this discussion.) Most culinary fats are composed of triglycerides made from a combination of different length fatty acids with different levels of saturation. When it comes to coconut oil, understanding the difference between the effects of medium-chain triglycerides (MCTs) and long-chain triglycerides (LCTs) is key, because although they only really differ in the length of their carbon chains, a small difference can greatly alter their effect on the body. It is also important to remember the difference between a fatty acid (a carbon chain with an acid on one end) and a triglyceride (three fatty acids stuck to a glycerol molecule). I appreciate that organic chemistry is not everyone's idea of fun, but bear with me. It will be worth it next time you find yourself in a health food shop.

Hydrogenated, refined, or virgin?

For use in baked goods and some sweets, vegetable oils are sometimes "partially hydrogenated," which involves adding some hydrogen to the carbon chains to turn unsaturated fats into

---

* Systematic reviews are generally the best kind of study, because they take all the available evidence from different experiments and consider whether it is possible to make a general conclusion. We shall discuss them in more detail in chapter 13.

saturated ones. This makes them more likely to form into a solid at room temperature and consequently useful in some industrial cooking applications. Unfortunately, this process can create a type of fat called trans fat, which has a different structure to other saturated fats and is known to be extremely harmful to your health. Consumption of large amounts in the diet has been shown to lead to an elevated risk of heart disease.[2] As a result, the use of hydrogenated fats has largely been phased out by food manufacturers, in favor of nonhydrogenated and often unsaturated alternatives. Although there is little doubt that trans fats from hydrogenated oils are particularly harmful, just because an oil has been refined, that doesn't mean it has been hydrogenated.

The key to coconut oil's reinvention as a superfood has been the development of the so-called virgin category of the oil. This is what David Wolfe consumes as medicine, Ella Woodward cooks with every day, and Gwyneth Paltrow . . . well, does whatever Gwyneth Paltrow does with it. Apparently, the miraculous healing properties of this cheap industrial cooking oil are revealed if it is made in a slightly different way. Refined coconut oils are produced in much the same way as other vegetable oils, with a lot of heating, bleaching, and deodorizing to produce a clear, stable, versatile oil. Whereas refined coconut oil is pressed from dried coconut and heated during processing, the virgin oil is made from fresh coconut flesh, by a process involving the same sort of cold pressing and extraction as you see with virgin olive oils.

The refining process of coconut oil does little to alter the fatty acid content. Refined and virgin oils have very similar compositions, but as with olive oils, some studies have shown a greater level of antioxidants and polyphenols (micronutrients thought to have some beneficial health effects) in the virgin oils.[3] Coconut oil, whether refined or virgin, is composed of around 45 percent lauric acid, a saturated fatty acid with a twelve-carbon long chain.* Lauric acid is by far the most prevalent fatty acid in coconut oil, and there is little doubt that it is functionally the most important.

MYTH 1: COCONUT OIL CAN CURE DISEASE

Although coconut oil has been linked to any number of health-giving properties and probably also crime-fighting superpowers, the main focus tends to be on its roles in disease prevention and weight loss. Both claims are idiotic and wrong, but for slightly different reasons.

First, let's focus on the myth of disease prevention. Lauric acid (the most prevalent fatty acid in coconut oil) is said to have powerful antimicrobial properties and the ability to fight numerous diseases. As the quote from Deliciously Ella shows, this sort of belief is not isolated to a few lone nut jobs. A quick internet search will reveal thousands more. How is it that this myth about what is essentially a saturated fat has drifted into the health lexicon?

Perhaps we should not be so quick to judge. Lauric acid is a remarkable and unusual fat and, believe it or not, has been shown to have antibacterial properties. Studies have shown it can kill a wide variety of potentially harmful microorganisms, so maybe some of these claims should be taken seriously.[4]

Or maybe not. One of the most common tactics of pseudoscience is to overplay the importance of specific scientific studies. Although the world of pseudoscience is keen to denigrate and dismiss scientific research when it comes to things like detox and alkaline foods, if a study seems to have any implication for their beliefs it will be accepted as gospel. The research (such as it is) into lauric acid

* It has about 16 percent myristic acid (fourteen carbons, saturated), 8 percent caprylic acid (eight carbons, saturated), and 7 percent capric acid (ten carbons). There are a few other bits, mostly longer-chain saturated fatty acids, and a low level of long-chain unsaturated ones.

entirely concerns the effect it has on microbes in a test tube, and absolutely no studies of note have shown lauric acid, or coconut oil, to have any effect on disease-causing pathogens in humans or even rainbow trout. Maybe the test-tube study sounds like an interesting lead, but the leap from test-tube effect to real-world health outcome is huge and all too common in pseudoscience circles (I am not sure they actually call them pseudoscience circles as that would give the game away, but you know what I mean). Other things that kill pathogenic microorganisms in a test tube include water, sugar, salt, dish soap, stamping on them, or setting them on fire, yet this information is not particularly useful when it comes to treating disease. In hospitals, the most effective way to kill almost all dangerous microorganisms before they enter the body is with an alcoholic hand gel, but that does not mean that consuming large quantities of alcohol will be of benefit to your health (although as a young chef I did spend a heroic amount of time performing this experiment, just to check). Lauric acid is no different. Although the effect was no doubt interesting to the scientists who investigated it, that does not mean that coconut oil is a miracle, cure-all superfood.

Perhaps we have a slightly romanticized view of scientific discovery: the lone genius in the laboratory making an earth-shattering breakthrough. The one example we all know is Alexander Fleming's discovery of penicillin in 1928, and yet even this was not quite the dramatic moment many of us might imagine. Fleming was a famously messy researcher who returned from vacation to find that one of his staphylococci cultures had been contaminated with mold. In the area around the contamination, the bacteria had stopped growing, leading Fleming to postulate that the mold must be producing some sort of antibacterial substance. His reaction to this discovery, perhaps one of the most profound and important developments of the twentieth century? Fleming is said to have remarked, "That's funny," before moving on to something else.

The fact that something kills bacteria in a dish is not enough to get any real scientists very excited. Although Fleming did eventually become interested in what this strange substance might be, when he published a paper on it the following year, no one paid much attention. He continued to work on it sporadically for a number of years before eventually giving up, as no one seemed interested. It wasn't until 1940, when Howard Florey and Ernst Chain started synthesizing it and testing it on animals, that the enormous potential of penicillin was realized, and by 1944 it was changing the course of the Second World War.

Although lauric acid is no penicillin, and no study has shown it to have an effect in the body, I suppose its ability to kill bacteria in a test tube might have some implications for the use of coconut oil as a mouthwash. Incredibly, this has been studied, albeit in an extremely limited way, showing a tiny effect on a very small group of people.[5] The bactericidal effect of oil pulling (a traditional Ayurvedic practice that involves swilling oil around your mouth for twenty minutes) was found to be comparable with that of a commercially available mouthwash, although the mouthwash works faster, is less unpleasant, and is less likely to result in you consuming excess amounts of saturated fat. Oh, and the experiment didn't use coconut oil because that is not what Ayurvedic practitioners use. But if you are ever stuck without mouthwash, have a couple of spoons of fat handy, and twenty minutes to spare, it might be worth a try. Other than raising your risk of coronary heart disease, what harm can it do? Vociferous coconut oil advocates will now be shouting at this book, desperate to point out that the antioxidants present in virgin coconut oil will also have an effect in fighting disease, thus justifying the enormous markups charged for this product. This may be true, and the presence of micronutrients in foods is often a good thing, but as we shall see in chapter 9, the case for antioxidants is far from clear.

MYTH 2: THE WEIGHT-LOSS MIRACLE

Far more ubiquitous than talk that coconut oil can fight disease is the belief that it is a miracle weight-loss ingredient. In many ways, although slightly counterintuitive, this myth makes more sense and has some corroborating evidence. It revolves around the effect of MCTs, which are thought to behave very differently in the body than LCTs. The smaller MCTs are absorbed and metabolized faster than LCTs, and, although the evidence is far from conclusive, it is increasingly thought that they may have a role to play in weight management. For instance, a recent study showed that 4 to 8 tablespoons (15 to 30 g) of MCTs has the potential to increase daily calorie expenditure by 5 percent, an average of about 120 calories per day.[6] This and other research has led to the frequent and much-repeated claim that coconut oil is actually a "fat-burning fat." The ability of MCTs to be quickly metabolized and mobilized as an energy source has also led to huge interest in their use in health, fitness, and endurance athletics circles.

MCTs are certainly interesting, and there is good research to back up the claims that they may have a role in weight management, which begs the question: What's wrong with the weight-loss claims of coconut oil advocates?

It's simple. Coconut oil actually contains very few MCTs. All the studies and clinical effects mentioned earlier look at oils that have been fractionated to remove all but the eight- to ten-carbon fatty acids (known as caprylic and capric acids). MCTs used in research and medicine do not contain lauric acid (twelve carbon), the main constituent of coconut oil. Most who study in this field do not consider lauric acid to be a MCT at all, and yet the effects of MCT experiments are frequently extrapolated to justify wild weight-loss claims for coconut oil.

The confusion is perhaps understandable. As an isolated chemical, lauric acid can be termed either a LCFA or a MCFA, but unfortunately, when it comes to biological function, it very much falls into the long-chain camp. This is the very reason that, when producing MCTs from coconut or palm oil for study or medical use, the lauric acid is removed. Lauric acid behaves entirely as a LCFA, with much lower solubility and slower rates of absorption and digestion. So, it is not a fat-burning fat. It is just a fat.

Additionally, even though coconut oil has around 15 percent MCFAs, when they are combined as triglycerides, they stick together in different ways, meaning that very few triglycerides (less than 4 percent) will actually fall into the MCT category—ones with a combined number of carbons between twenty-four and thirty.

To put the fat-burning claims into perspective, if coconut oil contains 4 percent MCTs and you want to benefit from the 5 percent increase of calorie expenditure you get from eating 7 to 8 tablespoons (30 g), you would have to eat 187 tablespoons (750 g) of coconut oil. Although you would burn an extra 120 calories in a day, you would be consuming over 6,000 calories of coconut fat to get there and would probably be sick.

In short, coconut oil should definitely not be taken as a weight-loss supplement. Anyone eating a spoonful of coconut oil is eating a spoonful of mostly long-chain saturated fat, a highly calorie-dense food proven to raise the risk of heart disease. So, the second anyone mentions MCT benefits in relation to coconut oil, you can now tell them to fuck off and learn a bit of organic chemistry. I told you that bit of chemistry in your life would be worth it.

So, what else can't it do?

There are no studies to show any disease-fighting effects of coconut oil in humans or animals. There are plenty of studies on metabolic effects relating to MCTs, but these have absolutely no relevance. Is there anything left?

Maybe one or two things. Coconut oil has been studied a little, as has its main fatty acid component, lauric acid. To say the results justify miracle-superfood status would be stretching things a little, but I guess as "miracle-superfood status" is a made-up marketing term, evidence is not really an issue there. One of the main pro–coconut oil arguments centers on the health and vitality of a number of indigenous populations around the world who consume a lot of it. Bruce Fife is particularly keen on stories of these healthy population—free from disease, staggeringly beautiful, flawless creatures from an Edenic paradise. Of course, he attributes every inch of their vitality to their consumption of his magical oil, but then, he also claims it can cure AIDS and cancer, so take his opinions as you wish. Arguments about the legendary health of Polynesian island dwellers are rife in the coconut-oil debate, especially regarding their heart health despite their eating a relatively high-fat diet.

There have been a number of population studies, but little can be taken from them that gives any support to pro–coconut oil arguments. There is little doubt that some indigenous Polynesian populations who eat lots of coconut have low levels of heart disease, but I hope you will have spotted the mixing of correlation and causation, and the potential for a number of blindingly obvious confounding factors, not least that these studies tend to be on populations that are highly active, with diets that are also full of fruits, vegetables, and oily fish.[7] Also, another quite significant point is that none of these populations actually consumes coconut oil, as it is a modern manufactured food completely unavailable to them. They may eat large amounts of coconut flesh, and some will even make the flesh into milk, but none extracts the oil. Although the fat is consumed, there is little doubt that consuming it as a whole food is likely to alter the way it is metabolized. It is the difference between eating sunflower seeds and drinking a bottle of sunflower oil.

Polynesian populations are not the only ones around the world to consume high levels of coconut. Sri Lankans, for instance, consume at least as much per head and have abnormally high levels of cardiovascular disease.[8] This does not prove anything either way about coconut, but science doesn't work when you only pick the evidence that confirms what you already believed.

Other evidence on dietary effects is scant and inconclusive. Many claim that it raises the level of good cholesterol (the good one is HDL, for all of you who didn't look it up earlier), although most studies have shown that it raises the bad one (LDL) as well, thus increasing total cholesterol, thought to be a risk factor for heart disease.[9] The evidence tends to show that the cholesterol-raising effects of coconut are not as bad as butter, but worse than olive and sunflower. A study of the Malay people is often quoted as evidence that coconut oil can be useful in reducing waist circumference, but this was a small and limited study on twenty people that only showed a significant effect on the men in the group and did not show any effect on any other factors measured, including total body weight.[10] This is hardly a ringing endorsement of coconut oil's effectiveness and, if we take it literally, only applies to men. In your face, Paltrow.

WHY COCONUT OIL?

The rise of a once-reviled manufactured oil to become a weight-loss wonder and sexual lubricant to the stars is a curious phenomenon and one that shows our susceptibility to believe in anecdotes and make judgments based upon our values. Coconut oil is the most curious of all the superfoods, but also the most commercialized, with a number of brands emerging in the last few years, all selling virtually identical products. I have been told by some sensible and informed people that they believe there are influences at work from Big Coconut lobbying groups, spreading false rumors about the dangers of carbs and creating imaginary health claims to drive sales.

Appealing though the idea of an organization called Big Coconut is, I have no time for conspiracy theories. However, a number of high-profile health bloggers are openly named as "ambassadors" for various brands of expensive virgin coconut oil. These ambassadors will frequently talk about health claims for the product they represent. If you look on the packaging of the products or in any official advertising, you will see that there is no mention of any miraculous health claims, because to make health claims there without scientific evidence would be illegal.

But when someone talks about a product on social media, in a book, or during an interview, that does not count as advertising and so it is allowed, even when it is accompanied by a link to a website selling the product.

So, is coconut oil, as the packaging suggests, a small, obscure culinary fat with very limited use? Or, as some people suggest, is it a miracle healing oil with huge appeal and great commercial value? I am sure that the coconut oil manufacturers are quite happy when those representing their brands make unsubstantiated claims that might otherwise be illegal, and I am aware that the most influential stars will get decent money for this representation. The brands know that without these claims their product is a small, obscure culinary fat with very limited use, but with them, it is a miracle healing oil with huge appeal and great commercial value. I doubt that all the celebrity endorsements I have mentioned here are financially remunerated—the quality of celebrities they have recruited is far too high for small brands to afford—but there has been clever use of a number of high-quality influencers, people readily followed by those making judgments based on values. I do not believe in a conspiracy, but there is a money train rolling, fueled by false belief, and those profiting are not about to let it stop.

For those who use coconut oil medicinally, we can only conclude from the available evidence that the health benefits they attribute to it are imagined. For those who use it with such ubiquity in their cooking, I have to question their palates and culinary judgment. For me it has an overpowering taste that is only really suitable for certain highly spiced dishes. The fact that most health bloggers fail to notice that it makes everything taste like sunscreen is yet another reason their views on food should be ignored.

So, coconut oil, the most bizarre of the miracle superfoods. This leads us nicely to **Rule Number 6 in the Angry Chef's Guide to Spotting Bullshit in the World of Food**: *They will talk about superfoods—never trust anyone who tells you a particular food has magical powers.*

Just as a footnote, I would like to make it clear that I have no problem with coconut oil. If people want to use it in their cooking, that is fine: No food should be excluded, and no food should be classified as "healthy" or "unhealthy." My problem is with classifying it as a "superfood" and giving it magical powers. With food, the key is variety, and if you enjoy casseroles that taste like shampoo, by all means carry on. Just try olive or vegetable oils sometimes, or maybe even a bit of butter. The taste will be your reward.

Chapter Eight

THE PALEO DIET

———————————

O ne thing I love about the Paleo diet, and perhaps one of the
secrets to its great appeal, is that it takes a small piece of science that we all remember from school and makes up a story
about it being the solution to all the world's health problems. In
this way it is similar to the alkaline diet, which does the same with
some basic high school chemistry. It is as if a Hollywood script
writer has decided he needs to put a bit of science into a movie
plotline and is eager to make sure it is something an untrained
audience understands.

Mayor: Well done, Captain Science, you managed to reverse
the polarity of the acidity and saved the world from the obesity monster.

Captain Science: Thank you, Mayor, I couldn't have done it
without Angry Chef's help.

(I have always dreamed of an Angry Chef action-movie franchise.)

Everyone remembers a little bit about pH and acids from school chemistry lessons. We all know that acids are bad and corrosive, with pictures of nasty-looking skulls, crossbones, and half-dissolved hands on the bottle. With this highly limited knowledge alone, the alkaline diet makes sense. All foods are neatly broken down into two distinct categories, one good, one bad. If acids are bad, alkaline must be good, so eat alkaline things. (Please don't eat alkaline things, as this would be very dangerous indeed; I am just trying to make a point.)

The alkaline diet gives people the solution to all their health problems in terms of the little bit of chemistry they can recall. This is clearly an effective strategy, so if you can find any other areas of science that people vaguely remember from school and exploit them in a similar way, you are on to a pseudoscience winner. I imagine that at the headquarters of Big Avocado, in the secret underground lair where Gwyneth Paltrow concocts her plans, they regularly hold meetings that go a little like this:

Paltrow Science: Anyone got any ideas? We've done acid and alkaline. Is there anything else we remember from school chemistry?

Henchperson 1: Ummmm. Carbon chains?

Paltrow Science: *Good*. Pass that over to the coconut-oil department. They may have a use for it.

Henchperson 2: Something about moles?

Paltrow Science: That's biology, stupid. But it does give me an idea. Is there anything we can remember from school biology lessons?

Henchperson 1: Photosynthesis? Is there something about chlorophyll?

Paltrow Science: Good. I like it. Write that down. Chlorophyll produces oxygen, I think, so maybe we could say it oxygenates the blood.

Henchperson 2: Surely no one would believe that. How would it carry on making oxygen after it's been eaten? At the very least it needs light and . . .

Paltrow Science: You'd be surprised what people will believe. Anything else?

Henchperson 2: I've got one. How about evolution? Everyone remembers Darwin and the theory of evolution.

Paltrow Science: Mmmm. Maybe. Although it is just a "theory." What do we remember about evolution?

Henchperson 1: There was something to do with beagles eating finches, I think.

Paltrow Science: OK. Not great. Anything else?

Henchperson 2: It happens really slowly.

Paltrow Science: Ah. Interesting. Happens slowly. I think we might just be on to something.

THE PALEO DIET

"Eat Paleo" is the cry of many a muscle-bound fitness freak, keen to embrace the visceral, back-to-nature, commonsense approach of this most scientific of diet schemes, not to mention all the protein. Rejoice, because with this new diet, we can all tear gleefully at giant dinosaur drumsticks, eat wild-harvested berries, and live how man was meant to live, with endless piles of meat dripping in delicious fat, the very best that nature provides.

For those concerned with the toxicity of modern life, Paleo takes things to the extreme. When we talked about detox, I mentioned that some people like to hark back to a lost Eden, when things were pure and uncontaminated by modernity. Well, the followers of the Paleo diet firmly believe they have found that Eden, a point in human history when all was perfect. The late Paleolithic period, when brave, strong, vibrant early humans proudly walked across open plains, communing perfectly with their natural environment—lean, muscular, and with perfect posture; free from the ailments of modernity; free from disease, depression, toxic brain fog, dangerous chemicals, refined sugar, and evil corporations conspiring to make them sick. Men spent their days hunting for rich and bountiful game meats locked in exciting, deadly combat with magnificent wild beasts. Women gathered berries and roots; tended beautiful, natural, organic dwellings; reared children; and fashioned clothing from skins and natural fibers. The grateful tribe would marvel at the brave returning male hunters, and all would feast on rich, sumptuous meats to give them strength for their next mission. These were active people, in harmony and balance with the natural world, and they were masters of all they surveyed. It was a blissful and perfect point in human existence, lost forever by the march of progress, with only glimpses available to us in a few remaining indigenous hunter-gatherer tribes around the world.

Around ten thousand years ago, when agriculture developed during the Neolithic revolution, Paleo man was lost forever, tragically confined to the dustbin of history by the march of agriculture. Our eating habits changed dramatically in a short time, switching to a heavily grain-based diet, and we started consuming dairy and cultivated legumes, things that we are just not genetically adapted to do. As a species, we did all of our evolving in the golden age, when men were men and women wore bikinis made out of mammoth fur, meaning that we are not suited to the modern diet with its grains, refined oils, string cheese, and Pop-Tarts. Eating modern food has made us fat and sick, and it is all down to the bit of biology that we all remember from school: Evolution is slow, and progress is fast. We are stuck with Paleolithic genes and a modern agricultural diet.

This was all outlined by Walter Voegtlin in 1975 in his book *The Stone Age Diet*, and it chugged quietly along with a few enthusiastic supporters, including academics. Then, in 2001, charismatic advocate Dr. Loren Cordain published *The Paleo Diet*, which quickly became a bestseller, sending the popularity of the Paleo diet into the stratosphere. Since then, numerous other advocates have jumped on the Paleo bandwagon (which presumably has wheels made out of stone and is pulled by saber-toothed tigers), including research biochemist Robb Wolf and fitness celebrity Mark Sisson.

Wolf explains the premise of Paleo as follows: The Paleo Diet is the healthiest way you can eat because it is the ONLY nutritional approach that works with your genetics to help you stay lean, strong and energetic!

Sisson says of his "Primal Blueprint," his version of the Paleo Diet (a name that Cordain has trademarked): The Primal Blueprint isn't a fad diet or exercise program, but a sustainable set of nutrition, fitness, and lifestyle behaviors that align you with your genes' expectations so that you can enjoy superior health and wellness without any of the conventional struggle or sacrifice.

Speaking of the typical Paleolithic man, someone whom he seems to have remarkable insight into, Sisson says: A likeable fellow, really, who, incidentally, also has a charming family— a strong, resourceful wife and two healthy children . . . You see, by modern standards, he would be the pinnacle of physiological vigor. Picture a tall, strapping man: lean, ripped, agile, even big-brained (by modern comparison).

I'm sold. How can I eat Paleo?

The principles of the Paleo diet are actually quite simple. As Cordain says, "The Paleo Diet is based upon everyday, modern foods that mimic the food groups of our pre-agricultural, hunter-gatherer ancestors." It involves the rejection of all the foods that our Paleolithic ancestors would not have recognized and the largely unrestricted consumption of any that they would. This means that you can eat lots of meat (but you should try to stick to grass-fed meats only), fish and seafood, fruits, vegetables, eggs, nuts, seeds, and certain oils. You should favor organic produce and avoid grains (the devil's food), legumes (including peanuts), dairy, refined sugar, potatoes (no idea what the problem with potatoes is), processed foods, salt, and refined oils.

Among a number of Paleo enthusiasts, there is an adoption of the alkaline hypothesis to support their rejection of dairy. Curious, as the alkaline crowd consider meats to be acidic, so I am not quite sure how Paleo fanatics resolve the high meat consumption in their diet. Meat, offal, and other animal products form the mainstay of the Paleo diet, making it much higher in protein and fat than conventional dietary recommendations. The diet does advocate the consumption of lots of vegetables and fruits, so although it is low in carbohydrates, it is not on the extreme end of that spectrum.

Since Cordain published his book in 2001, the diet has shot to prominence, driven largely by its adoption in fitness and body-building communities and because of its instinctive appeal. While alkaline, detox, and clean eating appeal to a largely female demographic, Paleo is very much a male-dominated approach to diet. Not only is the hypothesized Paleolithic lifestyle likely to appeal to a certain retrograde misogyny—the muscular male hunter bravely wrestling bears, while the women tend the children and pick a few berries, it is also a diet that actively encourages men to eat a lot of meat. Like the Atkins diet before it, it appeals to masculine, carnal flesh-eating desires, the pleasurable and almost unrestricted consumption of meats and animal fat being central to its premise.

There are, of course, many female advocates, but Paleo is a largely male domain with online communities full of dramatic transformations and previously flabby men showing off their now ripped torsos and boundless energy. The world of Paleo is rich with social proof, and the meteoric rise of the trend has doubtless been driven by a corresponding increase in the power of social media. Many aspirational celebrities have advocated Paleo, including Jack Osbourne, Kobe Bryant, Matthew McConaughey, Megan Fox, and Uma Thurman. Unfortunately for the Paleo movement, Miley Cyrus is also known to be a fan.

THE PASSIONS OF PALEO

For many people around the world, the diet does seem to work (at least if their goal is some temporary weight loss; the evidence for it curing all modern diseases is unsurprisingly quite sketchy), and there are few food movements with such passionate supporters as Paleo. Many believe that this hidden truth, that evolution has not had time to adapt our metabolism to the dietary changes brought by the agricultural revolution, is the secret to the obesity crisis and many other modern diseases of affluence. Sugar conspiracies apart, for those who spend time debunking dietary myths there are few subjects more likely to create conflict than daring to criticize Paleo. When I started writing about pseudoscience in food, I was told by a number of other bloggers to be careful with the Paleo lot. They are sure to engage and are passionate that the diet works, that the premise is sound, and that it holds the key to solving the health crisis of the modern age. I was warned that the community includes serious academics and it would be best for everyone if I just steer clear. I will now cheerfully disregard that advice.

The Paleo diet makes for a very neat, simple story. Much like alkaline, there are sprinklings of sensible advice in there (eating lots of fruit and vegetables and getting lots of fiber in your diet are generally good things), but the premise on which the diet is based is garbage. It is built on idolizing the past, oversimplifying complex processes, and, most of all, a misunderstanding of how evolution works.

I will admit that, for some people, following the diet has positive effects, but as we shall discuss, examples of something working do not constitute proof of effectiveness. Many anecdotes stuck together do not make data and are certainly not evidence for the maladaptation of our genes.

THE PROBLEM WITH PALEO

If we believe in evolution (and I am presuming that most people reading this will be in the pro-evolution camp), then we must also believe that our existence is just a result of a series of random, undirected events. Even knowing this, it is hard to look at the complexity of life and not imagine it as the result of some sort of design. We shall discuss this further later on in the book, but no matter how logical and removed we are, all of us are inclined to believe in the hidden wisdom of nature. The natural world is so breathtakingly beautiful that it can seem perfect at times, and when we have to accept that this beauty is just the result of an undirected, uncontrolled series of random mutations, it can challenge our common sense to the core.

Nature is not perfect. It is not designed. Evolution is a random process that has no ultimate goal, and things are not designed to fit their environment perfectly. Because it has no destination, it never arrives. It is always meddling, interfering, and changing. Evolution does not create one perfect being at a single point in time and then chug along unaltered as the world changes around it. The thought that mankind has thrived for the last ten thousand years, out of step with some inherent genetic makeup, is absurd. It is something that could not have occurred because the process of evolution would not have allowed it to.

Professor Alice Roberts is an anatomist, anthropologist, and professor of public engagement in science at the University of Birmingham. As she explains, "The whole premise that you can go back to a point in our evolutionary history and say, 'This is when we were at our healthiest, when our food matched our physiology best,' is so fundamentally flawed. Evolution doesn't stop. Most of us in Europe can digest milk into adulthood—and that's certainly an adaptation which has emerged since the Neolithic and the advent of dairy farming."

The point about milk, something that can be digested into adulthood by many European and African populations, is largely ignored by Paleo advocates. Lactase persistence, the mutation that allows this, first occurred between ten and twenty thousand years ago, and in European populations is the result of a single mutation.[1] Here, evolution can be seen selecting for traits without the need for huge timescales. We evolved quickly because that mutation allowed us to fit better into the environment.

Humans are an extraordinarily adaptable species, with that adaptability to different climates, environments, and especially diets being crucial to our success. This was never more evident than within the Paleolithic period, a point in our history when we colonized huge parts of the world. To think that there was some sort of homogeneous diet that our species evolved to eat defies any logical sense. As Professor Roberts puts it:

> Exactly which Paleolithic diet is to be revered? How about the Siberian one, where you eat just reindeer and horse for nine months of the year? Humans are magnificently omnivorous—we can eat just about anything and survive just about anywhere. And if we want to understand the diet and lifestyle that's best for us today, we'd do better to look at modern physiological and health research than to hope the ancients have some mysterious message for us.

Paleolithic people ate what they could, adapting their behavior to suit their many different environments, and it is doubtful that they would recognize any of the ingredients available to us today. The vast majority of the fruits, vegetables, and livestock we now consume are the result of agriculture, carefully selected for particular traits and as different to their ancestral species as a Chihuahua is to a western wolf. And the unfortunate reality is that we know precious little about the actual Paleolithic diet. Mark Thomas, professor of evolutionary genetics at University College London, knows more about these ancestors than just about anyone:

There is only one archaeological site in the world that has given detailed and accurate data on diet in the Paleolithic. You can get information from modern hunter-gatherers, but these are largely confined to marginal environments such as the high arctic, equatorial rainforests and some desert communities, so are not necessarily representative. We just don't know what Paleolithic people ate and within the paleo diet community there is a huge amount of speculation, much of which is likely to be false. They have built up a romanticized picture of the period, and that picture is riven with a great deal of prejudice. It is speculation knitted together with absurd or cherry-picked molecular just-so stories. They dream away carbohydrate and yet to think that people survived that period on a low-carbohydrate diet makes little sense. The human body has a huge glucose requirement and the existence of many copies of the salivary amylase gene indicates that carbohydrate has always formed an important part of the diet for hundreds of thousands of years. When you do look at modern hunter-gatherers, they always know where the underground tubers are because they are an important source of carbohydrate.

It appears that carbohydrates were an important part of our diets long before agriculture, and there is also much evidence—both from our physical adaptations and microfossil data—that meat did not form a huge part of the diet of many populations during the Paleolithic period, and that even the evil grains were widely consumed for much of it.[2] Within serious academic communities, the Paleo diet advocates are not taken seriously at all. The premise on which they have created their plan is deeply flawed and utterly foolish. Which raises the question . . .

WHY HAS IT BECOME SO POPULAR?

Professor Thomas believes that "the primary motivations of the Paleo community are commercial and their appeal is centered on weight loss. If there is one thing we do know about real Paleolithic people it is that they would not have been interested in losing weight."

Paleo has persisted and thrived for a number of reasons, but one of the most important is its effectiveness when it comes to short-term weight loss. For many, following the diet is an incredibly effective way to shed excess pounds in the short term, so then why does it bother me so much?

The truth is that any diet that creates rules and restrictions will result in weight loss—at least in the short term. Paleo offers a romanticized false premise and a few academic advocates to give it some validity, but it is in essence the same as any other diet. Although it might seem to be counterintuitive to say "eat as much meat as you want and you will still lose weight," and seemingly is the unlocking of some great primal secret, the reality is that Paleo works because it is likely to result in the consumption of fewer calories. Meat, with all its protein and fat, is highly satiating and, for people who struggle with the eternal hunger of dieting, Paleo—or similar low-carb diets—can provide a useful strategy for weight loss. People do not lose weight because of a magical synchronization with the requirements of their genes; they lose weight for the same reason they do on the Atkins diet, the alkaline diet, a gluten-free diet, by clean eating, or by sticking to a raw-food vegan diet. Rules create restriction, and restriction makes you eat fewer calories. The only reason for the success of this particular restriction diet is the viral power of its made-up premise.

Paleo is nothing new. It is about as realistic as the Flintstones, and in accepting the misunderstanding of science that underlies it, there is a real danger of abandoning the tenets of reason. Once you reject the voices of real experts in favor of charismatic advocates with a prettier story, you leave yourself wide open to packs of pseudoscience wolves.

THE LOW-CARB CONUNDRUM

There is not much point in pulling apart the nutritional science of the Paleo diet because it is nothing new. There are some tweaks and different views, but it is a weight-loss diet as old as the hills. Many weight-loss regimens over the years have been successful with similar protocols. Cut carbs, especially wheat and refined grains, eat lots of protein and fat, and watch the pounds drop off. Whichever way this is packaged, the remarkable and unavoidable fact is that for many people, this sort of diet works and works well. Does this mean that within it, despite it being at odds with almost all dietary guidelines around the world, there is actually some sense? Even though the premise of Paleo is wrong, should we actually all be eating a low-carb, high-fat diet for maximum health?

The debate around this alone could easily be the subject of a whole new book, with a number of serious and intelligent people (including some dietitians) advocating low-carb, high-fat diets for weight loss. Many also advocate such diets for improved general health, claiming—perhaps counterintuitively for those of us raised in an era of saturated-fat health warnings—that high-fat diets protect against heart disease. Increasingly, those interested in dietary health have to nail colors to the mast as to which is the best diet to recommend, either LCHF (low carb, high fat) or the more traditional recommendations based on the perhaps misleadingly titled "Mediterranean diet," based on complex carbs, balanced amounts of lean protein, limited saturated-fat intake, and a focus on swapping saturated fats for polyunsaturated options such as olive oil.

Although I am just a poorly qualified chef whose opinion has little value in this debate, I tend to nail my colors to the

Mediterranean diet rather than to any of the LCHF options. I am not saying I am right—and Angry Chef followers may be interested to know that my anonymous (and far better qualified) collaborator Captain Science, although not a strong LCHF advocate, has differing views to me on many aspects of this debate—but I am making my case on this issue before we move on to some more enjoyable pseudoscience debunking.

First, despite numerous messages in the media, there is still plenty of evidence that high saturated-fat consumption has a link to cardiovascular disease, and most of the LCHF diets do contain large amounts of saturated fat.[3] Second, although the many positive weight-loss stories of LCHF do have some statistical evidence to support them, sometimes there is a danger of mistaking weight loss for health. Losing pounds is important to many, but it is not the ultimate goal for everyone, and for large numbers of people, it should not be encouraged at all. Restricting carbohydrates from your diet can have many negative consequences, including the potential for B vitamin deficiencies (whole grains are an important source) and a diet low in fiber. Also, at the risk of borrowing some of Paleo's premise, we are creatures whose metabolism is very much geared toward the consumption of carbohydrates, with massive metabolic requirements for glucose and a number of strongly selected genetic traits that show carbs have always been part of our diet. The alternative ways our body has of meeting its glucose needs are highly inefficient, perhaps explaining why limiting carbohydrate consumption is an effective weight-loss strategy. Placing the body under stress by limiting its intake of a particular nutrient, especially when it comes to the extremely low-carb diets (such as the ketogenic diet), is not likely to be an ideal state for long-term health.

One of the great tricks that low-carb diets have up their sleeve, and perhaps one of the reasons why so many people become so passionate about them, is that in the very early stages of the diet limiting carbohydrates is likely to lead the body to use up its stores of glycogen (the short-term reserves of carbohydrate in our muscles). Because of the way glycogen is stored, this also releases a great deal of water, meaning that in the very early stages of a low-carb diet, most people will experience dramatic weight-loss results, up to several pounds in the first week. Although this is just water loss, it is highly motivating for some. Unfortunately, it is a one-off benefit of the diet, which can lead to frustration in the following weeks and to some quite rapid water-based weight gain should carbohydrates be reintroduced. Perhaps quite tellingly, despite this initial burst of success, studies have shown that although LCHF can be effective, over a twelve-month period there is little difference between this and other, more conventional, lower-fat diet options.[4]

I believe that the main benefit of low-carb dieting is the satiating effect of the high levels of fat and protein—the ability they have to stop us from being constantly hungry when we are limiting our calorie intake. This leads many people to stick to this sort of diet rather than other more balanced and conventional options. Constant hunger is a big driver of diet failure for many, and a diet that seems to be unrestricted can be of great appeal. I suppose that given the conflicting and inconclusive information on whether LCHF is actually an unhealthy way to eat, the fact that a number of people use it as a weight-loss intervention should be welcomed. If it works for them, then, despite certain risks, this is probably fine.

My objection comes when this diet is held up as being the solution for all. There are a number of groups such as the Atkins company in the US and the Public Health Collaboration (PHC) in the UK that advocate this form of eating as the solution to all our dietary problems and that it should be the recommended choice for all.[5] Low-carb diets work for some as a weight-loss mechanism, but they are not the answer for everyone. Many people find them very difficult to stick to, partly because of a desire to eat carbohydrates, something that we are metabolically adapted to crave. They also come with a number of potential side effects, including a depressed immune system (perhaps explaining the much-reported "low-carb flu"), exhaustion, and, in the case of extreme ketogenic diets (where carbohydrate intake is severely limited so as to force the body to metabolize fat rather than the usual glucose), the potential in women for disrupted estrogen production, with a possible impact on bone health.[6]

Much LCHF support, especially in Paleo communities (but also in more supposedly scientific organizations), comes from anecdotal proof. There is a belief that if it "works for me," then it must be the solution for all. In understanding that correlation is not causation, we must also understand that large numbers of positive examples do not constitute proof that something works. A common battle cry among people who rail against pseudoscience is that "the plural of anecdote is not data." Paleo websites are full of anecdotal success stories of dramatic weight loss, but these stories are by their nature selective, especially in the sort of limited social media groupings we all exist in. For every low-carb, Paleo, or Atkins success story, there will be many unspoken failures. For people wanting to lose weight, the strategy that works for them is the most effective. The problem is when that strategy is held up as the solution to all our health problems.

This leads nicely to **Rule Number 7 in the Angry Chef's Guide to Spotting Bullshit in the World of Food**: *They will want you to believe that anecdotes are evidence.*

WTF IS WRONG WITH POTATOES

Although I am aware that this might expose my own bias as a chef and food lover, restrictive diets that demonize particular ingredients always sit uncomfortably with me, especially when the demon is something as important to our functioning as carbohydrates. Although the standard dietary advice and the Mediterranean diet do shame saturated fat to an extent, at least it is delivered as a message of balance, of making some healthier choices and trying to meet people where they are. Unfortunately, that sounds sort of boring compared to solutions that promise dramatic results, so let me tell you instead one of my own personal reasons why I wouldn't cut out carbs from my diet: potatoes.

Despite being a completely natural whole food, potatoes are arbitrarily excluded from the Paleo diet and universally demonized by low-carb advocates. Potatoes are delicious, versatile, cheap, and accessible. From a culinary and scientific point of view they are deeply fascinating and they form the basis of many of my favorite foods. Cooked correctly, they can make the heart sing with joy. I believe that I know how to make the perfect roast potatoes, crisped to the point of near caramelization on the outside, with a light fluffy interior, a hint of thyme, and the richness of beef fat. I can make mashed potatoes so silky, rich, buttery, and creamy that they once made someone cry. I once spent an entire week at work trying to make the perfect fries (they are good, but not perfect—yet). I love dauphinoise, lyonnaise, röstis, sautéed, Bombay, chips, waffles, hash browns, Hasselback, croquettes, gnocchi, and patatas bravas. I love shepherd's pies and hot pots, and believe that potatoes slowly cooked with meats often become more delicious than the meat itself. When made well, freshly cooked fries are as much of a cause for celebration as the flashy, needy offerings of Michelin-starred restaurants and superstar chefs. When eaten out of paper by the sea, they are the greatest culinary pleasure I know.

In all my years working in professional kitchens, I have learned to spot a good chef when I see one, but it is only when I see someone work with potatoes that I truly know if they can cook. For potatoes need care, skill, precision, timing, refinement, and an understanding of their science to really make them sing. They are a joy, the embodiment of everything I love about food. They are the humblest of ingredients, but with a little time, knowledge, and skill they can be transformed into something sublime.

During the harsh winters of the Second World War, when meat was scarce and most fruits and vegetables were unavailable, potatoes nourished and fueled Britain's war effort, forming an integral part of the Dig for Victory campaign (Note: I wanted to claim here that potatoes won us the war, but apparently in this edition I have to acknowledge some US involvement). They were nutrient dense, loved by all, versatile, cheap, and delicious. If you cannot fit potatoes into a healthy balanced diet, there is something wrong with the way you are eating. They are a near perfect food, and to create rules that exclude them not only feels wrong, it is wrong.

Arbitrary exclusion of perfectly healthy ingredients such as potatoes in such diets as Paleo exposes their true nature. They are the pursuit of thinness hidden behind a veil of health, all justified by oversimplification and misunderstandings of science. Potatoes expose this perfectly, which only makes me love them all the more.

Chapter Nine

ANTIOXIDANTS

——————————

HEROES AND VILLAINS

I really love tea.

Sometimes I drink seven or eight cups a day. Large mug, nice and strong, dash of milk. It is an elixir that can bring me to life in the morning and keeps me going throughout the afternoon. Enough of it will cure a hangover, soothe an illness, and improve any situation. It fuels and hydrates every day of my life. I pity countries and societies that do not embrace tea as enthusiastically as the British. They don't know what they are missing; it is the finest drink known to man.

Incredibly, in a rare positive twist of fate, tea has near magical health-giving properties. It is packed full of antioxidants and, as we know, antioxidants are miraculous bringers of good health. They are the cure-all superagents of nutrition, repelling into your body like a highly trained SWAT team to rid it of villainous free radicals. Free radicals are vile wrongdoers, electron thieves that make violent raids into your cells. They crash around causing all sorts of distress and havoc, ripping electrons from your DNA, smashing up lipid membranes, and even helping to trap LDL (the famous bad cholesterol), making it stick to artery walls.

Often the result of exposure to radiation, environmental toxins, and cigarette smoke, free radicals are a vicious gang of terrorists unleashed into your body. They have been linked to cancer, vision loss, memory problems, heart disease, strokes, and a number of other chronic conditions. Just when we think all is lost, in come the magical antioxidants. Each one has a specific mission to rid the body of a violent attacker. They have heroic names like vitamin C, vitamin E, beta-carotene, and coenzyme Q10. They swoop in, grab a free radical, neutralize it, and take it out of harm's way. The Chuck Norris of nutrients. Even more powerful are selenium and manganese, capable of forming the active site of superpowerful antioxidant enzymes that can take out large numbers of free radicals. Like John Rambo with an M16.

If I have overindulged a bit at lunchtime, which is entirely possible when the most attractive-looking option in the staff canteen is deep-fried, I will have a cup of tea or two in the afternoon. It might sound a bit weird, but when I drink it, I can almost feel the antioxidant SWAT team plunging down into my body to do its work, ridding me of any harm caused by my fried indulgence. I feel the same when I have my first cup of tea the morning after a heavy night. My head might be pounding, my stomach may churn, but I have a crack team of elite special forces antioxidants to bring me back to life and heal my damaged body. It is science knowledge in action, and it makes me love tea all the more.

Sometimes things just work out well. We enjoy something, and it does us nothing but good. Antioxidants are a great example. Tea, red wine, chocolate: They are all rich in these hugely beneficial nutrients. Thank you, science, you have made life good. Pour the wine, crack open that bar of chocolate, and pop the kettle on. Antioxidants are the definition of a superingredient, guaranteed to protect us from the toxins of modern living.

Or are they?

But antioxidants are an unstoppable health panacea, right? And tea is a wonder drink that cures all ills.

I mean they are, aren't they? I am not just making up this stuff. (Science gives me a withering look.)

Oh really, science. Are you going to take this one away from me, too? Sometimes it feels that everything I think I know is wrong. That'll teach me for getting all my information from newspaper headlines.

PALTROW SCIENCE

For a moment I want you to imagine a dystopian world where Gwyneth Paltrow has been put in charge of science.

Gwyneth and her health-blogger brethren are fans of cognitive ease. They like certainty. I imagine that Gwyneth would soon retire Science Columbo from service (remember him from chapter 6?). If something had a nice story and provided an elegant solution to a problem, it would be enshrined in law. Science would be clear and constant and would have a defined, simple message. It would be broken down into bite-size blocks of certainty that we could all relate to. To define the law of universal gravitation in everyone's mind as an everlasting truth, it would be featured as Goop's "Theory of the Week," with a special note of endorsement from Gwyneth: *OMG, Isaac is sooo our fave physicist here at Goop and we just love his new law of universal gravitation—it's really simple and fits into our busy lifestyles.* Dissenters who challenged it would be quashed with a vicious program of juice cleanses and mindful wellness videos.

Would it matter? The law of universal gravitation would probably not be the worst thing that Gwyneth has endorsed (remember what she does with coconut oil). It works pretty well. We could probably get a spacecraft to the moon using it (although in my new dystopian world, it might have to be a holistic spacecraft powered by healing crystals). But with no Science Columbo, no one would come up with a theory of general relativity. As a result, we would have far less understanding of the universe. We would never have reached a theory of black holes or an understanding of the Big Bang. Worse still, we would still be balancing maps on our knees while driving, as our GPS navigation systems depend on measurements so sensitive they are affected by general relativity. Also, modern advances in charged particle therapy treatments for cancer would be beyond our grasp as the huge speeds involved mean that the behaviors of particles are predicted using Einstein's breakthrough.

Without a constant process for challenging even the ideas that are pretty good, there would be no scientific progress. Science Columbo is responsible for some of mankind's greatest advances. His existence within the framework of the scientific method gives science the ability to see past simple, easy-to-believe narratives and constantly search for the truth.

SCIENCE COLUMBO AND THE ANTIOXIDANTS

So, to antioxidants. As I have discussed, there was a simple, easily understood story that brave superhero antioxidants help eradicate villainous free radicals. During the early 1990s, this theory was communicated widely, based mostly on in vitro observations of animal cells. As a recent biochemistry graduate at that time, I thoroughly bought into it, as did pretty much everyone else. Experimental evidence was supported by lots of data showing that people with a diet high in antioxidant-rich foods—fruits, vegetables, that sort of thing—had a lower incidence of cancer and numerous other chronic conditions. It all seemed fairly promising, and the world's media communicated the message with gusto. All hail antioxidants, the perfect natural antidote to the poisons of modern life.

I think we can imagine how Science Columbo approached the matter.

Science Columbo: There's just one thing that's bothering me.

Scientists: What's that?

Science Columbo: All these results look promising. But how does it perform in the real world? Can we get some evidence that these "antioxidants" work in the human body?

Scientists: Of course we can. You just leave that to us, Science Columbo.

And off the scientists went. Experiments were designed, often involving antioxidant supplementation to gauge the size of the effects on chronic disease. Just how good were antioxidants at protecting people against the negative health effects of free radicals? If the theories bore out, there should have been significant effects on the incidence of cancer, heart disease, vision loss, and strokes.

Ah. Bit of a problem. The results in the real world tended to range from poor to downright disastrous. Almost all the trials conducted on antioxidant supplementation had massively disappointing results. Sometimes a bit more than just disappointing. Meta-analysis* of supplement users revealed that people taking beta-carotene, vitamin C, and high-dose vitamin E actually had increased rates of mortality.[1] A large US study showed that beta carotene and vitamin A increased the risk of lung cancer.[2] A Finnish trial on beta-carotene supplementation in smokers had to be

---

* A meta-analysis is the statistical analysis of a number of scientific studies in a particular area, combining data to gain more robust results.

called off when there was shown to be a significant rise in cancer incidence.[3]

In all, there is precious little evidence to suggest that antioxidant supplements have any beneficial outcomes for chronic disease. That is not to say that antioxidants in the diet do not have a role, but when you are studying the food people eat, it is nearly impossible to attach a causal link to specific nutrients. Just because people eating a diet rich in antioxidants show improved health outcomes, we cannot assume that antioxidants cause the improvements. It should not really come as a surprise that someone eating a diet with plenty of fruit and vegetables might have lower incidence of chronic disease than someone eating hardly any. As we have learned, correlation is not always causation, and when it comes to dietary changes, there are so many confounding factors that drawing conclusions about a single nutrient should be avoided, however nice the story. We ingest hundreds of different chemicals every day, and it is extraordinarily difficult to work out the effects of just one. If you want a summary of the evidence supporting the effects of antioxidants in our diet, it would be "we have no idea."

Maybe we should not be so surprised. Although laboratory testing of animal cells in vitro did show potentially beneficial effects, the reality has proved to be a lot more complicated. It appears that free radicals and antioxidants both play important roles in regulating key systems in the body. Importantly, free radicals might not be quite the villains they have been made out to be.

For instance, free radicals are now generally thought to play an important part in the mechanisms of immune function. White blood cells have been shown to release them when attacking bacteria. It also seems that the free radicals produced during exercise are responsible for many of the associated health benefits, and that trying to counter them with the high-dose antioxidants that many people recommend might be extremely counterproductive. There is even some experimental evidence that antioxidants, our all-conquering heroes, might help to keep cancer cells alive and assist in metastasis (the spread of cancer through the body that leads to many cancer deaths).[4]

The picture is hazy at best. Antioxidants should not be thought of as a single group of hero substances with magical healing powers. They should not really be thought of as a group of substances at all, as the effects of antioxidants can vary widely. Some of the substances we know as antioxidants, such as vitamins C and E, can actually have a free radical effect at high doses. Each of the many hundreds of substances that can exhibit antioxidant properties is different and is likely to perform a different role in the maintenance of the body. Much as in life, there is no good versus evil. Perhaps life would be simpler if there was, but the reality is murky, complex, and highly nuanced. As usual, the best advice is not to focus too much on one single group of nutrients and to try and achieve a bit of balance.

What now?

Antioxidants are not an all-conquering health panacea. Science Columbo has done his work, so clearly the media will have reported these facts comprehensively, and we can all move on. Or so it should have done. Unlike with universal gravitation, the disappointing message about antioxidants has somehow not managed to filter through to the public. If it isn't true (or at least not nearly as true as we used to think), how can it be that the majority of people still believe in it? From a personal point of view, despite being reasonably well informed, until recently I firmly believed that antioxidants in my daily cups of tea were healing my body. I thought the science was proven. Like most people, I did not feel the need to research this belief before I proclaimed it. I did not check the efficacy of every study conducted. I did not quote references. Like most people, I believed it was true because I vaguely remembered reading it somewhere and my instinctive brain had decided that it sounded about right. Tea does feel nice.

This process of blind acceptance based on previous behavior is sometimes known as self-herding. It is similar to the better known concept of pack behavior, that we are influenced by the behaviors of others in group settings, but here the social proof comes from memories of the way we have acted in the past. I was completely comfortable in my belief because it sounded correct and I had believed it in the past. It did not matter that I could not remember why I knew it or where I had heard it first. I happily stood in line, nodding in agreement with my previously believing selves, never really doubting why I was standing there. I felt this so strongly that I imagined brave antioxidant heroes seeping through my body every time I drank a cup of tea.

THE EXTENT OF OUR IGNORANCE

Given that I spent so long holding onto this false belief, despite my science background and naturally skeptical nature, it is perhaps not surprising that the magical power of antioxidants is widely popular in the world of clean-eating health bloggers.

Take this small selection from some of my favorite clean-eating websites:

From Ella Woodward: Coconuts . . . also contain lovely antioxidants that help to protect our bodies from disease!

From Kimberley Snyder: If your skin is feeling a little dry and dull this time of year, no worries! This is actually the perfect time to load up on brightening and immunity-boosting antioxidants. My new **Skin-Glowing**

Antioxidant Smoothie Recipe is here to save the day, your skin and your immune system!

From the Hemsleys, on *Harper's Bazaar*: Watercress is bursting with antioxidants. It's excellent for combatting late nights as well as helping you recover post-gym session.

From an interview with Adam Cunliffe on Goop: The high antioxidant content of many plant-based foods is protective of cells at the level of DNA, mopping up dangerous free radicals that can damage our genes.

From Mark Sisson in *The Primal Blueprint*: Antioxidants serve as a powerful first line of defence against oxidative damage from aging, stress, and inflammation. Moreover, antioxidants appear to contain cancer-fighting properties and to support the immune system (among many other benefits).

Or this from various small health-food brands:

From Alex Jones of Infowars: The heart and brain need appropriate amounts of antioxidants on a daily basis in an effort to support against constant toxin exposure.

From beauty products company KYPRIS: Antioxidant Dew is full of antioxidants and vitamins to protect your skin from free radical damage.

From Goop, on the Intense Nurture Antioxidant Elixir: A super-serum that acts as a protective cocoon for skin.

From Clean and Lean on its Body Brilliance supplement: Contains powerful antioxidants (like tasty blueberries and goji berries) which can help your body fight free radicals.

Trust me, this is just a small selection. The list goes on. It appears that many people have failed to read the memo on antioxidants. Or at least they have chosen not to believe it if they did. It is a slightly different story in the world of big food and drink manufacturing. In 2009 Tetley said the following in an advertisement for green tea: "For an easy way to help look after yourself, pick up Tetley Green Tea. It's full of antioxidants." Shortly afterward, Tetley was forced to withdraw this statement by the Advertising Standards Authority in the UK because the company had implied "that the tea had health benefits beyond hydration, in particular because it contained antioxidants" and that they "did not hold substantiation for the implied claims."

PALTROW SCIENCE LIVES

Even though claims on retail products are not allowed, it appears that my parallel world, where Gwyneth Paltrow is in charge of science, is not too far from the truth when it comes to antioxidants. Although the real-world benefits vary from nonexistent to vanishingly small, their power is still hailed by many. When it comes to the ever-popular antioxidant supplements, the current evidence is incredibly weak and in many cases quite damning. Somehow, though, the antioxidant train rolls on. Antioxidant supplement sales are worth around $500 million per annum in the United States, despite the fact that there are no proven health benefits from taking them when it comes to improving the symptoms of chronic diseases. Health bloggers, small manufacturers, self-styled nutritionists, and gurus all extol the virtues of antioxidant-rich foods and supplements, endowing them with magical properties. When a person's livelihood depends on a particular belief, it is often understandably hard to convince them that it is not true.

Does it matter that public knowledge is a long way behind the scientific consensus on antioxidants? I think it does. Aside from the fact that many people are spending huge amounts on supplements for no good reason, misinformation can have dangerous consequences. When New York–based cancer patients were surveyed a few years ago, it was found that around 60 percent took antioxidant supplements during their treatment, often at high doses and without the knowledge of their physician. Remember that studies have shown antioxidant supplements can have a negative effect on the success of cancer treatments. Somehow, though, patients remain so convinced by the antioxidant story that they are willing to disregard the evidence. This is the real problem with ignoring bad science. Once you let the door open just a crack, you are in danger of letting in the flood.

I still love tea. I still drink seven or eight cups every day. I have been given some small comfort by a study on flavonoids (that likewise target free radicals) in tea and chocolate that showed them to be protective against cardiovascular disease.[5] The study has not been proven by randomized controlled trials yet, but it makes for a good story, so I am hanging on to that one (another withering look from science).

Apart from that, I now accept the work of Science Columbo and have moved on from any fantasies of magical antioxidant powers. I am reluctantly shifting my thinking to the mixed, complex, and nuanced picture common to so many areas of science, and have bid a sad goodbye to my Chuck Norris–like antioxidant heroes. It's a shame, but that is the cost of scientific progress—to let go of existing ideas in search of the truth. It's hard to do, but if we want the world to move forward, we need to try to persuade others to do the same.

Chapter Ten

SUGAR*

If the world of action movies has taught us anything, it is that when something is imbued with magical superpowers, it needs to have an equally powerful archenemy. Although there are a few candidates for a potential nemesis to our brave superfood heroes, these days the Lex Luthor of bullshit nutrition is our old friend sugar. In a bizarre and unexpected twist, much like that bit in *Terminator 2* when we discover Arnold Schwarzenegger is actually on the side of good, the once-reviled saturated fat has been revealed as our friend. The true enemy is refined carbohydrate, a dangerous shapeshifting symbol of evil, hiding in everything we eat.

Although there are a million different gurus, armed with online certificates, personal health journeys, and differing views on how to holistically detoxify the body from within, the one thing they all agree on is that sugar is bad. Worse than bad, it is a vile, toxic poison, violating our bodies, causing us to be obese, depressed,

---

As this chapter is about sugar, it is an unwritten rule that I must include a number of terrible sugar-related puns in the subheadings, so apologies in advance.

diseased, and broken. Sugar is all that is wrong with the world, its consumption driven by corporate greed and conceived in the minds of corrupt scientists. It is sweet poison, more addictive than crack cocaine, more destructive than crystal meth, a silent killer and blight on the lives of us all.

If you think I am overstating, here are some fun quotes:

From Madeline "Get the Glow" Shaw: This white powder is now known as the perpetrator behind diseases such as diabetes and our growing obesity epidemic. The problem is that sugar is now in everything we eat, often masquerading behind names such as glucose or different syrups . . . There's a lot of research which suggests that sugar is a highly addictive substance . . . French research-ers have reported that in recent tests, their laboratory rats chose sugar over cocaine [. . .] even though the rats were in fact addicted to cocaine!

From Mercola.com, the website of Dr. Joseph Mercola: Sugar is loaded into your soft drinks, fruit juices, and sports drinks, and hidden in almost all processed foods—from bologna to pretzels to Worcestershire sauce to cheese spread. And now most infant formula has the sugar equivalent of one can of Coca-Cola, so babies are being metabolically poisoned from day one.

From Natural News: The entire scope of drug addiction has been observed in people with sugar addiction. There are cravings, an escalation of tolerance levels, and dramatic withdrawal symptoms associated with sugar addiction that parallel that [sic] of both prescription and non-prescription "street" drugs.

From Goop: Sugar gives you an initial high, then you crash, then you crave more, so you consume more sugar. It's this series of highs and lows that provoke unnecessary stress on your adrenals. You get anxious, moody (sugar is a mood-altering drug) and eventually you feel exhausted. Sugar is also associated with many chronic problems that include decreased immunity, some chronic infections, autoimmune diseases, heart disease, diabetes, pain syndromes, irritable bowel syndrome, ADD, chronic fatigue, and candida.

From poisoning babies to making the world obese, we are left in no doubt. Sugar is a toxic, addictive poison, maybe even the sole cause of all our health problems. We are the addicted slaves of a malevolent, despicable food manufacturing industry. Within the debate on sugar, there is anger, outrage, and disgust. There are evil villains hell-bent on our destruction and brave warriors sent to avenge us. The antisugar saviors promise to guide us through our detox, break our cycle of addiction, and expose the greedy and corrupt institutions that have damaged us for commercial gain.

Although, in the interests of balance, sugar does appear to have some use in getting rats off cocaine, so there's a silver lining to every cloud . . .

A STICKY SITUATION

I am going to level with you here. Sugar is not completely in the clear. Most of us eat too much of it. The US Office of Disease Prevention and Health Promotion recommends that we get no more than 10 percent of our calories from "added sugar" so as to maintain a healthy balanced diet, while it is perhaps worth noting that the World Health Organization recommends a lower target of 5 percent. Although it is very difficult to get a true picture of how much sugar we actually consume (especially how much added sugar), a National Center for Health Statistics survey published in 2013 estimated that the average consumption by adults in the US was around 13 percent. The picture is quite a lot worse for children and adolescents, who are thought to get a genuinely shocking 16 percent of their calories from added sugar.[1]

Too much sugar in our diet can cause a number of health problems. It certainly increases the likelihood of tooth decay, especially when consumed as a snack between meals. Also, although the science is nowhere near as definitive as many would like to claim, a high added-sugar diet can lead to putting on weight and all of the associated health problems. Sugar is a facilitator of excess calorie consumption, with many high-sugar products being very palatable ways of consuming a lot of excess fuel, but that alone does not make it the cause of obesity.

I have no problem with sensible, balanced calls for people to cut down on their sugar intake. Given that we all eat too much of it, and there are a number of genuine health problems that might result from this excess, I think that such calls are completely justified. I love food and want people to have a healthy relationship with what they eat, and however nice fizzy cola bottles are, anyone consuming 15 percent of their daily calories from added sugar is unlikely to be embracing the rich diversity of ingredients available to us.

The problem I have with the debate around sugar is twofold. First, the pseudoscience, misunderstandings, and conspiracy theories that surround it just confuse people and don't help to address the problems of obesity at all. Second, and perhaps more distressingly, the language of guilt and shame drives me to despair. For this chapter, I want to deal with these problems. Sugar is a subject that causes me much pain—like many who long for a world in which we have a sensible, realistic, and balanced relationship with our food. If you want to see a frustrated middle-aged chef raging at the world, this is the chapter for you.

THE SWEET CONSPIRACY

There is a much-repeated conspiracy theory about sugar, born from a belief that powerful lobbying groups, the mysterious people of Big Sugar, have been colluding with scientists since the 1960s to set dietary guidelines so as to drive sales. In the United States during the 1950s and 1960s, there was a tremendous rise in the incidence of coronary heart disease. A number of scientists, led by the charismatic nutritionist Ancel Keys, attributed this to an increase in the saturated fat consumption of the average American diet. After much research and debate, in 1980 the US government released dietary guidelines advising people to cut back on foods containing saturated fats and cholesterol in an attempt to combat this growing health crisis. In 1983, the UK government took similar action to combat its own problems.

Although these measures were issued with the intention of improving public health, many people think they may have had a dramatic and unintended consequence. Data show that although US obesity levels remained fairly flat until 1980, shortly after the new guidelines were issued, obesity rates began to rise sharply, a trend that has continued almost unabated ever since. We can see a similar story in the UK, which now has some of the highest rates of obesity in Europe. A number of high-profile campaigners have postulated, many quite insistently, that the change in the dietary guidelines directly led to the rise in obesity. (At this point I would like to remind you about the dangers of mistaking correlation for causation.) More than that, the blame is often put firmly on sugar, because food manufacturers seeking low-fat formulations to meet new consumer demand often replaced fat with sugar to maintain palatability. Refined starches and sugars replaced dairy fats, butter was cut from people's diets in favor of sweetened low-fat spreads, and high-sugar breakfast cereals took the place of more traditional fatty options, such as eggs and bacon.

Campaigners argue that these changes to the dietary advice made the world fat, and did so because they drove huge increases in the levels of sugar consumption. For them the evidence is clear. In 1980 the guidelines were changed, advising people to lower their saturated fat intake. The food manufacturing industry planned this, with lobbying groups bribing scientists and governments, desperate to force the public to consume more sugar. Everyone switched to a high-sugar, low-fat diet. Everyone got fat. The conspiracy continues because scientists do not want to admit they have been wrong, and Big Sugar still has the world of nutritional science and public health bodies firmly in its power. We are told that overwhelming scientific evidence in favor of fat and against sugar is being ignored and that the dietary advice being issued around the world is fundamentally flawed.

This story is backed by a number of academics and campaigners (Robert Lustig, David Gillespie, Zoë Harcombe, and Aseem Malhotra, among them) who cite a great deal of evidence for the dietary evils of sugar and fructose. Many label it as a toxin, something that we are not metabolically adapted to consume, and say that weight gain is not to do with an excess consumption of calories, but more about the metabolic damage that our vastly increased consumption of sugar has caused. Many of these campaigners also claim that saturated fat has been wrongly vilified.

I am not desperately keen to go into every detail of the science here, but in my opinion the correlation between the rise in obesity rates and the changes to the dietary guidelines is a near-perfect example of a hare sitting next to a pile of eggs.[2] What is more, beyond the assumption of causation, a long and complex narrative has been created about conspiracies, evil scientists, and good versus bad foods.

THE EVIDENCE GOES SOUR

Did the dietary guidelines issued in 1980 cause obesity? Perhaps they were a factor, but to make any sort of causal link to a single nutrient when it comes to a problem as huge and complex as obesity is too simplistic. In truth, the guidelines issued at that time were along the lines of "Eat a more balanced diet with lots of fruit and vegetables and plenty of fiber." The guidelines actually recommended cutting down on sugar as well as fat, but it is true to say that they were communicated to the public with a great deal of focus on the "fat is bad" message. Fat went on to be demonized quite vociferously by the media and diet gurus of the time. Fat, and particularly saturated fat, was blamed for all our health problems by many, falling for a convenient, simplistic narrative and neglecting to understand the real nature of the advice being issued. People are never interested in subtle changes to make small improvements in their health—they are far more interested in big changes and revolutionary improvements, such as fat-free options. Removing fat from the diet became the only goal for many, neglecting everything else.

Although there is plenty of anecdotal evidence of a change in people's diets and a raft of lower-fat options that were made available to consumers throughout the 1980s and 1990s, one of the most damning pieces of evidence against the sugar conspiracy comes from the distinct lack of evidence that the change in guidelines actually caused much shift in sugar consumption. Although the United States did see a per capita increase, and sugar consumption in the US shows a strong correlation to obesity levels, the so-called Australian Paradox (referring to a paper published in 2011 that showed declines in Australian sugar consumption had coincided with increases in rates of obesity) provides some evidence that in certain populations, sugar consumption actually declined after 1980, but where this occurred, obesity rates still increased.[3] Although consumption data are notoriously unreliable and the Australian Paradox has been the subject of much accusation and debate, it does seem to indicate at the very least that the causes of obesity are more complex and multifactored than a simple increase in the consumption of one nutrient. In the UK, the Department for the Environment, Farming, and Rural Affairs (DEFRA) has carried out annual surveys on the UK diet, using diet journals and marketplace receipts, and has found plenty of evidence that sugar consumption is declining. Since 1992, despite a continuing rise in obesity levels, these surveys showed a per capita fall in sugar consumption of 16 percent.[4] English data also show that between 2002 and 2012, there was a 7.4 percent fall in sugar consumption accompanied by a two-kilogram rise in the average adult body weight.[5] According to a British Heart Foundation report in 2012: "Overall intake of calories, fat and saturated fat has decreased since the 1970s. This trend is accompanied by a decrease in sugar and salt intake and an increase in fibre and vegetable intake."[6]

For many people this might be a little surprising, given it comes from such a respected body in a serious report and completely counters the conspiracy-theory rhetoric more often heard in the media. Passions run high, and the authors of "The Australian Paradox" received much vitriol, being accused (and then cleared) of academic malpractice. The very fact that even the authors of the paper considered it to be a "paradox" goes to show how strongly held the beliefs are that sugar consumption and obesity are essentially the same problem. Although there is evidence that the two are correlated in the United States, surely the lack of supporting data from other countries is enough to make us wonder whether there is a confounding factor. Mostly this evidence is ignored, and those reporting it are vilified, yet no one has offered to explain the counterevidence, either.

Sugar campaigners assume that consumption dramatically increased all around the world after 1980 because it fits well with their theory and with personal anecdotes. Many of us of a certain age have fond and distant memories of drinking delicious, high-fat milk and eating thickly buttered bacon sandwiches, yet this belief of a golden age is not based upon detailed food journals, just on a notion that things were better "back then." We imagine a magical time before the world became corrupt and broken, before the "scientists" started telling us what to eat with their guidelines and controls. A time when we all lived in harmony and ate the perfect natural diet. The reality is not so clear. I am old enough to remember a time before the guidelines were issued—and sugary drinks, candy, and doughnuts were not invented in 1980.

The lack of per capita consumption evidence supporting the sugar conspiracy hypothesis is not definitive proof that there is no link. (Or as Joseph Heller once put it, "Just because you're paranoid doesn't mean they aren't after you.") Obesity happens to individuals, not whole populations, and rises in obesity do not occur uniformly. Perhaps if there were overwhelming scientific evidence of a causal link between the two, that would be more persuasive. In 2014, the international journal *Advances in Nutrition* published a review of all the available evidence on the link between sugar and obesity and found that "current research trials conducted with commonly consumed sugars do not support a unique relation to obesity, metabolic syndrome, diabetes, risk factors for heart disease, or nonalcoholic fatty liver disease."[7] When looking specifically at obesity, they combined the data from three recent systematic reviews of sugar consumption and body weight and found that "taken together, these meta-analyses of RCTs [randomized controlled trials] demonstrate that replacing sugar with other energy-equivalent macronutrients has no effect on body weight." So, any causal link between the two should at least be open to debate.

As anyone working in public health will tell you, the real problem with dietary guidelines is not the advice; the problem is that no one follows it.[8] If the cause of obesity was as simple as a change in the guidelines, then we could probably stop worrying about it. Many health bodies around the world have recently reviewed sugar advice, leading to a large reduction in the recommended amounts and different strategies to decrease consumption. If the public are such slaves to government advice, then this will surely slash sugar consumption in half, yet I fear the reality is quite different (cue antisugar campaigners' shouting that this is only because we have become so addicted in the last thirty years).

SUGARCOATING THE TRUTH?

Let's say we are the sugar conspiracists starting from scratch in the 1960s. Let's look at the practicalities. We are determined to make the world stop eating saturated fats, knowing in advance that they will be replaced with huge quantities of sugar. We will have to be so powerful we can persuade the nutritional establishment to spend millions of dollars producing fake science to justify these new dietary guidelines. We will then have to bribe every level of government to enforce these guidelines on an unsuspecting public. Then we have to continue to ensure they are enforced, paying off the medical establishment, the WHO, numerous charities, public health bodies, and nutrition researchers around the world, and keep producing systematic reviews that show links between consumption of saturated fats and increased risk of heart disease.

And who exactly is funding this operation? Who has to gain from false guidelines being issued? The mysterious forces of Big Sugar? Maybe, but in all honesty, who is more powerful in the world of agriculture, with the ability to influence policy: Big Sugar or the enormous might of US beef and dairy industries, not to mention the fast-food restaurants selling high-saturated-fat products by the millions?

The guidelines issued in the 1980s were the first ones that actually suggested people should restrict their consumption in some way. Are we really supposed to think that food manufacturers would have a vested interest in dietary guidelines telling people to eat less? Would the food manufacturing industry want to hide information that showed fat consumption has no link to ill health, and that high-fat products can be eaten with impunity?

As anyone who works with food will know, the key to creating high levels of palatability is not either sugar or fat; it is a combination of the two. It is that delicious point where sweetness and richness combine in chocolate, cakes, icings, doughnuts, and sweetened cream that makes them almost irresistible delicacies. These are the sources of calories we are most likely to consume in huge amounts with little nutritional payoff. Few of us over the age of six would eat a tablespoonful of sugar on its own. Few people could stomach eating a packet of butter or drinking a pint of cream. But mix the sweetness of sugar and the richness of fat together, and you have a perfect combination for driving food sales. That is what a nefarious food manufacturing industry with full control of nutritional guidelines would demand freedom to sell.

If saturated fat had no link to heart disease, many powerful people within the food industry would have a vested interest in sharing that information with the world. The idea of the food manufacturing industry fixing dietary guidelines to force an increased consumption of sugar is one of the most absurd conspiracy theories of our time, and yet it is one of the most accepted and readily believed. Here are a few examples:

In a House of Lords debate, speaking about the way the body digests fat, the Conservative Peer Lord McColl of Dulwich: This is a beautifully balanced mechanism, which tends to prevent us eating too much and thereby prevents obesity.

Not surprisingly, the food industry does not approve of this beautifully balanced mechanism, because it has resulted in less food being eaten and lower profits. So it joined up with the some [sic] rather dubious scientists to produce research that erroneously claimed to show that fat was bad and carbohydrates were good.

From Mercola.com: In short, federal dietary recommendations have very little to do with actual nutrition science, and everything to do with promoting foods that serve the junk food industry's bottom line, not the public health.

From a synopsis of *The Sugar Conspiracy*: This compelling investigative documentary exposes the US sugar industry's systematic hijacking of scientific study to bury evidence that sugar is, in fact, toxic. For forty years, Big Sugar deflected threats to its multibillion-dollar empire . . .

We fear obesity, a vast and highly conspicuous health crisis, and with that fear comes outrage. We desperately want someone to blame. We want clear explanations and simple solutions. The terrifying truth is that obesity has landed upon the world. No one really knows why, and no one can make it go away. It is a deep structural problem and, without returning to a dark age of food deprivation and shortage, it might even be one that we cannot solve. With our innate desire to make sense of the world, this is about the hardest thing possible for us to accept.

We desire control and crave explanation—they may even be more addictive than cocaine—and even the brightest and best will latch on to far-fetched stories that take away a little bit of the frightening unknown. If we can blame a malevolent food industry, then at least someone is in control, which is preferable to the unpalatable truth that no one is in control at all.

BITTER USE OF LANGUAGE

Sugar is not a great dietary evil, and although we eat too much of it and many of us would benefit from cutting down, it is demonized to an extent that is at best unhelpful and at worst irresponsible. I get annoyed about conspiracy theories, and frustrated that certain campaigners are given more media attention than is justified, but when it comes to the language of shaming, there is little that upsets me more.

In a world struggling with obesity, sometimes it seems that it has become acceptable for the gloves to come off. People can say whatever they wish so long as their intention is to "cure" diet-related health problems. Or to put it another way, lying, shaming, and prejudice are all fine as long as you are directing your hatred toward fat people.

The language used to describe sugar in the mainstream media is deeply shocking. We are regularly told it is a drug, an evil white powder, profoundly toxic with no amount of consumption being safe. To free ourselves from enslavement, we must be led through detox by our avenging sugar-free gurus. Be warned, we will go through cold turkey if we do, we will know the desolation of the addicted soul, we will shake and crave, but when we have escaped this demon's vile grip, we will be free. Free from the foul poison, free from corporate and government conspiracy. The promised land of perfect health and vitality awaits those with the strength for the battle.

Here are some book titles from the last couple of years. Books like these dominate the health and diet bestseller lists, with their authors leading the charge against this most pernicious of enemies:

- *No Sugar Diet: A Complete No Sugar Diet Book, 7 Day Sugar Detox for Beginners, Recipes & How to Quit Sugar Cravings*
- *Zero Sugar Diet: The 14-Day Plan to Flatten Your Belly, Crush Cravings, and Help Keep You Lean for Life*
- *Sugar Detox for Beginners: A Quick Start Guide to Bust Sugar Cravings, Stop Sugar Addiction, Increase Energy and Lose Weight with the Sugar Detox Diet*
- *Sugar Busters!: Cut Sugar to Trim Fat*
- *I Quit Sugar: Your Complete 8-Week Detox Program and Cookbook*

Apparently you get bonus points if you also mention the word *detox*.

We all eat sugar every day. You cannot go sugar-free. Every diet contains a certain amount of sugar, because sugar is contained within every fruit, vegetable, and grain and in all dairy. Sugar is not a poison, nor a toxin, nor a drug. In sensible amounts it can be part of a balanced diet. If you want to quit sugar completely, you will have to consume only fat and proteins, and you will get very ill.

Not one of the diets that claims to be sugar-free is actually free from sugar. No one who claims to have quit sugar has done anything of the sort. They have cut down on sugar, and I am happy for them, as I firmly believe most of us could do with cutting down, but they have not quit. And if you think this is just semantics and doesn't really matter, I am going to tell you exactly why it does.

Many of the diets I have mentioned here advocate the use of clever "natural sugar replacers," such as honey, date syrup, maple syrup, agave syrup, or similar. Some of the new sugar-free gurus (I am looking at you, Sarah Wilson) actually sell some of these natural sugar replacers on their websites. These are all effective substitutions of sugar because they contain sugar. Lots of it. They are mostly sugar. The concept of a natural sugar replacer is in reality quite bizarre because sugar is of course natural, extracted from plants without any chemical modification. Refining does not imbue it with some sort of toxicity and does not make it any more or less harmful than the same chemicals in a slightly different form and a pretty-looking bottle.

All fruits contain sugars that are chemically very similar to the refined sugar that you buy from the supermarket, but when it is held in the fiber and pectin matrix of whole fruit, it is released into the body more slowly, meaning that it affects the body differently. When that matrix is broken, however, for example in any of the popular sugar substitutes or when making a juice or smoothie, all that benefit is lost.

Sugar is just sugar. It is not good or bad. It is only toxic if you eat too much of it, and by that measure all food is toxic (remember the dose makes the poison: you can die from drinking too much water). When sugar is swirling around inside you, your body doesn't care how expensive the bottle was.

To attach the language of toxicity and addiction to a food we literally cannot avoid means pouring vitriol and scorn on all of us. To give a child a bowl of breakfast cereal and a glass of orange juice is now to be a social pariah, guilty of the vilest abuse, guilty of poisoning your child. If you allow your child to have some sweets, something that many loving and nurturing parents might well do on occasion, you are left feeling that you may as well be rubbing crack cocaine under their eyelids. If you are not sugar-free, if your children are not sugar-free, then you are toxic, you are contaminated. You and your kin are foul and pestilent junkies, no longer worthy, no longer deserving of medical treatment, contaminated and beyond hope, destined to die fat, lonely, miserable, and addicted.

This is the language of shame, and it is attached to something that in sensible quantities does no harm. Even in the dark days of fat demonization, there was nothing this extreme. The media is free to humiliate the fat, to blame them for their ignorance and stupidity. To denigrate people for their physical characteristics and attribute negative character traits is the very definition of prejudice. In any other circumstances, this would be unacceptable, and yet weight shaming—the vilification of fat and the belief that it implies moral weakness—has become commonplace. This is out of fucking hand, and it needs to stop.

THE BITTEREST IRONY

What is even more incredible is the sheer dumb-fuck irony of the whole sugar conspiracy argument. "Look," they say, "how stupid were we back in the 1980s to point our fingers at saturated fat as the great dietary demon? How idiotic to blame that alone, how foolish to think that it caused heart disease and ill health? How did we not see the dangers of blaming just saturated fat? We were shortsighted and blind, with reductionist arguments and language that singled out a lone nutrient as the cause of a complex problem.

"Now we have moved on and are far more sensible! Now, with all the advances we have made and all the lessons we have learned, we realize that sugar is responsible for all our problems! Sugar alone is making us fat and causing our ill health. We have finally singled out a lone nutrient as the cause of a complex problem."

And while all this is going on, they will say that science is broken. They will claim that it is hardly surprising that scientists are not trusted because they keep on changing their mind about what we should and should not eat.

Do these people not realize that we are in danger of lurching from one unbalanced diet to another? Whatever the conspiracy theorists might say, sound nutritional advice has changed little over the years. The answer has always been available. Here it is, the Holy Grail, the answer to The Great Question of What We Should Eat . . .

Actually, I think I might leave that secret until later on, just in case you all stop reading now and go off and live perfect lives with shiny, beautiful hair. For now, let's just say that we shouldn't spend our lives lurching from one great evil to the next, searching for the secret to set us free from an imaginary demon. There is no secret; there is no demon. It's all just food. If it was really, properly harmful, it would be illegal to sell it as food—like bleach, for instance. Never drink bleach.

Sugar is important. It is the sweetness of freshly picked peas. It is in the bliss of strawberries warmed by the sun. The sweetness of foods can enrich our lives; it can bring us joy and pleasure. Sweetness is a vital part of the flavor palette of every cook and true sweetness comes only from sugar. From a pinch to balance the acidity of a sauce, to the dark indulgence of a rich caramel, sugar has a place in increasing the enjoyment and palatability of foods, enriching our relationship with what we eat, and enhancing a lifetime of culinary joy. To dismiss and reject the use of sugar because of false notions of toxicity is to misunderstand what it is to eat healthily. A healthy diet is one of joy, not one of rejection and denial. If you want to eat for health, you should have nothing to do with the creation of arbitrary, senseless rules and certainly should never feel the slightest guilt over an occasional indulgence.

In essence, nutritional advice has never really changed. Serious scientists are not arguing about whether saturated fat or sugar is bad, because the reality is that neither of them is. Neither is evil; neither is toxic. Neither should be excluded, and yet neither should be eaten with impunity. The message has always been the same. You can and should eat them both. Just not too much. And lots of other stuff, too.

Part III

THE INFLUENCE OF

PSEUDOSCIENCE

Chapter Eleven

A HISTORY OF QUACKS

In trying to understand the rise of the most recent food fads and pseudoscientific beliefs, it makes sense to look a little into the past. Over the past few hundred years many different strange, illogical beliefs have evolved around food and health—hardly surprising given the ubiquity of these two things in all societies. Over this time a number of health gurus have come and gone, and although many vanish without a trace, the influence of some remains to this day.

It is important to remember when dipping into the rich history of dietary pseudoscience that in eras before the advent of modern medical treatments, there was perhaps even more inclination for people to believe the simple narratives of quackery. It may well have been that the fads and snake oils were the best option people had, and although treatments sometimes did more harm than good, when disease strikes, there is always the desire to intervene in some way, even if that intervention is ultimately futile.

THE DOCTRINE OF SIGNATURES

Perhaps the first profoundly mistaken guru of food health on record is seventeenth-century botanist William Coles. Influenced by the writings of German mystic Jacob Boehme, he popularized the doctrine of signatures, an idea that foods, and especially medicinal herbs, display characteristics that provide a visual clue as to their potential health-giving properties. This was rooted in religious beliefs, an idea that God would want to show mankind what use his gifts to the world might bring. A sort of divine game of medical Pictionary.

Walnuts were thought to help with diseases of the brain; the stinkhorn mushroom (*Phallus impudicus*), with male fertility; and eyebright, a plant whose flowers supposedly resemble a pair of bright blue eyes, was thought to treat eye infections. Coles reasoned that:

> Though Sin and Satan have plunged mankinde into an Ocean of Infirmities, yet the Mercy of God, which is over all his workes, maketh Grasse to grow upon the Mountaines and Herbes for the use of men, and hath not only stamped upon them a distinct forme but also hath given them particular Signatures whereby a man may read the use of them.

This is clearly very silly and wrong. Perhaps the greatest indictment of how silly and wrong it is comes from none other than Samuel Hahnemann, the founding father of homeopathy,*

---

* Homeopathy is a system of "medicine" created by Hahnemann in the late 1700s. Preparations to treat disease are made by performing repeated dilutions of a chosen substance, taking care to bash the container against a leather-bound book at each stage. The dilutions are repeated until none of the original substance remains in the solution, leaving only distilled water, which is then

ironically describing it as "the folly of ancient physicians." When the crown prince of all quackery denounces you as nothing but a quack, you know you are probably on shaky ground. To be fair, a number of homeopathic remedies are based upon plants used by William Coles, so although it might be folly to use an extract of the plant itself, clearly Hahnemann felt that if you dilute it down until there is nothing left but water, then the effect can be explained.

Similar belief systems to the doctrine of signatures exist across many cultures and are thought to be a relic of a prewritten language era, used as a way of creating visual associations to help remember the medicinal use of certain plants.[1] These associations were likely made after any medicinal use was discovered, a sort of meme preservation system from a time before guidebooks or search engines. Over the years it somehow became mixed with essentialism and religious belief, and what began as a simple memory aid ended up as a proof of divine guidance. (It is funny how food, health, and religion are so often intertwined.)

Many relics of William Coles' ideas remain today, with such foods as mussels and oysters imbued with aphrodisiac qualities because of a resemblance to female genitalia (apparently). Perhaps slightly unbelievably, the doctrine of signatures is undergoing something of a modern renaissance, with a number of new age herbalist and naturopathic websites detailing the power of this principle in defining the medicinal properties of various foods.

Georgian quacks and vegetarians
Within the Georgian era (1714–1830), the doctrine of signatures waned somewhat, perhaps because of the introduction of many New-World

used to treat conditions. These beliefs persist to this day, and modern homeopaths believe that the water has a memory of the substance once dissolved in it, and that the memory is activated when you hit the water with the book. In comparison, many of William Coles' ideas seem quite sensible.

ingredients, such as potatoes and corn, that did not fit into the existing belief system. After all, potatoes only really look like potatoes. It is also possible that such magical thinking may have been thought to belong to a previous age, as the ideas of science were starting to influence the world. The last English monarch to practice the "king's evil," the belief that a touch from the monarch would cure disease, was Queen Anne, who died in 1714, heralding the beginning of the Enlightenment and the end of certain ancient mysticisms.

Annoyingly, Georgian times do seem to have begun the great obsession with "personal health journeys," and there are many journals of the time showing a fanatical interest in bowel movements and weight. In some respects, very little has changed, and in historical terms, we should count ourselves lucky that internet blogs are a recent invention.

The Georgian era's great dietary quack was the infamous Scottish doctor George Cheyne, in many ways the founding father of self-help nutritional quackery. A friend and compatriot of Isaac Newton, Cheyne is perhaps the first in history to fit the patented Angry Chef Health Blogger Template. In his influential book *The English Malady*, he wrote:

> Upon my coming to London, I all of a sudden changed my whole manner of living . . . my health was in a few years brought into great distress, by so sudden and violent a change. I grew excessively fat, short-breathed, lethargic and listless . . . About this time, my legs broke out all over in scorbutic ulcers, the ichor of which corroded the very skin where it lay any time and the fore parts of both legs were one continued sore.

Enjoying the high life in London, he became obese, at one point reaching 32 stone (450 pounds), before a health crisis drove him to a life of abstinence and moderation, crucially removing meat from his diet. He lost a dramatic amount of weight and began to preach a message of meat-free moderation to his many rich and influential patients. He also wrote a number of books on the subject, with a preaching style and language not unlike modern-day clean-eating bloggers. In his work *An Essay of Health and Long Life*, he writes that his diet is "designed for persons who are able and willing to abstain from everything hurtful, and to deny themselves anything their appetites craved, to conform to any rules for a tolerable degree of health, ease, and freedom of spirits."

In an echo of modern-day healthy eating fads, not only did he credit his diet with weight loss, he also believed that it had a remarkable ability to cure disease. In *The English Malady*, he attributes his "Vegetable and Milk Diet" with remarkable curative properties (perhaps the one thing he could have learned from the modern day is a slightly catchier name for his diet):

> There are some cases wherein a vegetable and milk diet seems absolutely necessary, as in severe and habitual gouts, rheumatisms, cancerous, leprous, and scrofulous disorders; extreme nervous colics, epilepsies, violent hysteric fits, melancholy, consumptions (and the like disorders, mentioned in the preface), and towards the last stages of all chronic distempers. In such distempers I have seldom seen such a diet fail of a good effect at last.

Tellingly, the "English malady" that he refers to in his most important work is not obesity or any of the other physical afflictions that he claimed his diet could cure. Cheyne believed that his diet's most important power was in curing mental illness, believing that obesity was due to melancholy, caused by the luxury and affluence of a modern urban lifestyle. In his later work *Natural*

Method of Curing the Diseases of the Body, and the Disorders of the Mind Depending on the Body, he writes that his diet is "the true and real antidote and preservative from wrong-headedness, irregular and disorderly intellect and functions, from loss of the rational faculties, memory, and senses."

Cheyne is a bizarre and complex character, uncannily reminiscent of the modern health guru (although I have no evidence that he married a painfully dull musician or gave his children ridiculous names). He was so wrapped up in a sheltered, elitist existence that he lost touch with any sense of reality. His idea that obesity was the great malady of the time is in stark contrast to the grinding poverty and deprivation that most of the urban population was forced to endure. For all but a privileged few, meat was a rare luxury, and to cut it from the diet for health or moral reasons was unthinkable. The real English malady was social inequality and unsanitary urban living. Cheyne's musings are as ridiculous and out of touch as the latest insta-celebrity complaining of exhaustion, imagined intolerances, and brain fog.

Cheyne's diet is perhaps one of the first to advocate the exclusion of particular foods for health reasons, and although this was only possible because of his wealth and luxury, it is interesting to see the germination of this trend. Cheyne openly speaks of a moral aversion to the consumption of meat, with little attempt to hide links to religious abstinence and purity. This is in contrast to the health gurus of today, who repeat many of the tenets of Cheyne's regimes, but keep their moralizing quite implicit. The far-fetched and made-up health claims are much the same, however, as is the belief in anecdote as evidence and the evangelical zeal developed from personal experience.

Most of all, Cheyne's writings were some of the first to connect food and health with moral values, especially when linked to vegetarianism, something that would continue through the Victorian era and to the present day. He came from a life of privilege, felt inspired by his recovery from a personal health crisis, and signaled his moral virtue to the world by abstinence, denial, and a rejection of meat. Sound like anyone you know?

Victorians, experts on quackery

As we enter the Victorian era, quackery for the masses becomes possible. The urbanization of populations, the availability of newspapers and advertising, and an awareness of the power of science and technology drove it into a dangerous new era.

Brands started to have a powerful influence, with a number of deeply unpleasant and harmful syrups or cordials available for sale. Godfrey's Cordial and Mrs. Winslow's Soothing Syrup were both opium-based treatments for babies and children, known as "the poor child's nurse" as they were a cheaper and more effective way of stopping the cries of a hungry child than food. Many children died of starvation as a result of these remedies, a rarely told scandal of the time. It is hard to know how many children lost their lives as victims of opium addiction, as most deaths were recorded as starvation or malnutrition, but the use of these dangerous treatments was widespread throughout the nineteenth century.

Other bizarre patent medicines flourished, with wild claims that they would cure a huge variety of maladies. The miraculous Daffy's Elixir, originally developed by clergyman Anthony Daffy, was perhaps the most widely sold of all. Essentially little more than a laxative dissolved in gin, it was marketed with the promise of extraordinary curative powers for a wide variety of diseases, a common tactic for purveyors of quackery keen to ensure as large a potential market as possible. Daffy's was marketed as a treatment for gout, scurvy, gripe, colic, distemper, rickets, wind, and many other ailments. Its use on babies and small children was especially common—they would tend to quiet down once they had had enough gin.

All the while vegetarianism flourished in isolated pockets, a movement defined by political activism and a rejection of the establishment. Many religious groups, socialists, and feminists rejected the newly developed middling cuisine of Victorian Britain, preferring instead a meat-free rejection of the orthodoxy. Much of this movement was driven by a romantic idolization of the preindustrialized past, and a distrust of the modern urbanized society that had developed.

Although rejection of meat was largely political, false associations with health developed, too. The new vegetarian campaigners attributed contamination and disease to meat, saying it was filled with parasites, that it transmitted infections and that its consumption caused cancer and gout. Here we see the origins of moralistic pseudoscience, untruths told to reinforce a narrative. Vegetarian campaigners formed a number of societies in the Victorian era and would regularly challenge meat-eating competitors to sporting contests, keen to prove their vitality and strength. Disillusioned with urban living, groups of late nineteenth-century intellectuals saw preindustrialized peasant-style living as an ideal, ignoring the grinding poverty, disease, and malnutrition that had characterized this lost paradise.

As well as meat, white bread was demonized by many campaigners, mostly for political and religious reasons. For many, the democratization of white bread, once the rarefied choice of the elite, was a symbol of a broken world. Campaigners longed for the return to more traditional whole-grain grinding methods and a simpler way of life, safe from their fear of modernity.

At this time, there was much to be fearful about. Cities were poorly sanitized; cholera, typhoid, and gastric diseases were incredibly common; and the food supply was rife with the threat of contamination. Much of the flour sold was adulterated with chalk, plaster of paris, or alum so as to meet consumers' insatiable demand. Beer and milk were routinely watered down, and contamination from unscrupulous or ignorant manufacturers were extremely common. In 1858 a Bradford sweets maker called Joseph Neal accidentally coated a batch of mint humbugs (a hard candy popular in the UK) in arsenic, which he had bought thinking it was a cheap food adulterant called gypsum. Hundreds fell ill and twenty-one people died, leading to public outcry and the eventual introduction of regulation to prevent the adulteration of foodstuffs.

The new urbanized poor of the Victorian era were completely dependent upon manufactured foods and unable to afford much meat, so the rejection of any food was still the preserve of a rich intellectual elite. Contaminations and scares did lead to the creation of well-known brands, largely developed to assure an increasingly concerned public. Crosse & Blackwell promised preserved goods free from copper contamination. Tea brands such as Tetley grew from guarantees of an unadulterated product. Nestlé grew by selling uncontaminated, preserved milk. It was a desire for safety and purity that allowed large food manufacturers to grow from the urban sprawl of Victorian England.

The strange puritanical health fads of the time also drove the creation of many significant and lasting food businesses. The same social reformers who rejected the orthodoxy of the new middling cuisine also proved to be remarkable culinary innovators. Many chocolate brands such as Fry's, Cadbury, and Rowntree were formed by British Quakers as part of the temperance movement, selling cocoa as a substitute for alcoholic drinks (and also promising a product free from brick-dust contamination). In the United States, vegetarians from the Seventh-Day Adventist movement developed Kellogg's Corn Flakes to encourage the consumption of breakfasts based on whole grains. Another Seventh-Day Adventist, Sylvester Graham, founded a movement based on beliefs in vegetarianism and the consumption of whole grains, inspiring the creation of the "graham cracker" that is still produced today. Looking at attitudes to the food industry today, it is strange to think that many of the largest brands and manufacturers were born out of either a desire for purity, or a moralistic rejection of the modern world.

THE QUACK-BAITER GENERAL

As for the quack medicines, that industry flourished throughout the Victorian era, and in doing so, it developed many of the tricks and techniques still used by pseudoscience today: the fake testimonials, the celebrity endorsements, the wild claims of cure-all ubiquity, the power of advertising and media.

In 1905 the industry came under the scrutiny of the journalist Samuel Hopkins Adams in a series of articles for *Collier's Weekly* entitled "The Great American Fraud." In them, he brilliantly exposed this industry to the world, showing its dangers and revealing the dirty tricks used to mislead the public. Such was Adams' influence and the power of his writing that action was taken, with the Pure Food and Drug Act passed just a year later. Adams was perhaps the first great quack-buster, exposing an issue of great danger to public health and campaigning for change to protect people from harm. One cannot help but think that if he were around today, there are a few battles he would enjoy fighting.

"The Great American Fraud" is fascinating to read now, as it exposes practices that will be familiar to anyone interested in modern quackery. Although the Victorian panaceas and snake oils have disappeared from shelves, the same language and techniques are used to sell diet books, detox smoothies, herbal treatments, homeopathy, and many other quack therapies. One of my favorite sections reads:

> There it is in a nutshell; the faith cure. Not the stimulant, but the faith inspired by the advertisement and encouraged

by the stimulant does the work—or seems to do it. If the public drugger can convince his patron that she is well, she *is* well—for his purposes. In the case of such diseases as naturally tend to cure themselves, no greater harm is done than the parting of a fool and his money.

He also talks of the dangers faced by those seeking a patent cure for more serious ailments. "The most devilish of all [are those that] destroy hope where hope is struggling against bitter odds for existence."

Adams is a quack-fighting hero, but sadly, while Edwardian-era patent medicines are a distant memory, many of the practices revealed in "The Great American Fraud" are relevant now. The regulation may tighten, but pseudoscience and quackery will always grow in the cracks, capable of taking over if allowed to flourish unchecked.

Learning from history

To look back at the last few hundred years is to see the development of many fads and misunderstandings that persist to this day. We see claims of universal cures, the idolization of times past, the needless classification of foods as good or bad, status (or class) signaling, and the medicalization of our diets. Despite enormous progress, little has changed, and little has been learned. Regulation in some areas means that those who would once have pushed Daffy's Elixir or Graham's Cordial now sell life-changing detox diet plans. Reformers once moralized about the need to eat grains; now they moralize about not eating grains. The brands that once signaled safety and purity are now rejected for the opposite reasons, with small manufacturers seen as the embodiment of all that is pure. The quacks still aim to deceive and, for as long as they keep it up, the quack-baiters will rail against them.

Chapter Twelve

THE POWER OF ANCIENT WISDOM

I don't go in for ancient wisdom. I don't believe just 'cause ideas are tenacious it means they're worthy.

—Tim Minchin, "White Wine in the Sun"

All your toxins come out on your tongue, so you want to remove them. I use a tongue scraper. You can buy them from Indian Ayurvedic shops.

—Jasmine Hemsley

Never does the wellness industry seem more confused than when it cites ancient wisdoms to prove its case. Hippocrates is frequently quoted on healthy-eating websites to support the validity of many an outlandish claim. Popular Hippocratic quotes include "Let food be thy medicine and medicine be thy food" and "All disease begins in the gut."

Within the wellness community, a quote from the father of modern medicine is held as an undeniable truth. Hippocrates also said, "Eunuchs do not take gout, nor become bald" and "A physician without a knowledge of astronomy has no right to call himself a physician." But it was a very different time, and our understanding of the universe has developed considerably since then. Despite his undoubted influence on modern medical practice and a body of work that moved forward the treatment of disease, Hippocrates was ignorant and wrong about many things. Just because an ancient quote can be found to support whatever woo you might be peddling, it does not constitute an endorsement from the world of modern medicine.

This brings us to **Rule Number 8 in the Angry Chef's Guide to Spotting Bullshit in the World of Food**, and the subject of this chapter: *They will quote ancient wisdoms and expect you to accept them as fact.*

Unfortunately, a lack of evidence does not stop many ancient wisdoms from exerting a strong influence on the health and wellness industry. Hippocrates aside, wisdoms from ancient Greece are rarely considered, and the many archaic medical belief systems linked to Christianity and Islam are ignored, too, perhaps for slightly different reasons.

As we saw in the case of Robert Young and his attraction to pleomorphism, much modern-day nutritional woo can be traced back to splits that occurred in the medical community around the turn of the twentieth century. I would like to consider two considerably more ancient belief systems that strongly influence alternative medical and nutrition communities and have done so for many years. In this chapter we look at traditional Chinese medicine and Ayurvedic practices to see why they are so persistent and what they can teach us about more recent health fads.

TRADITIONAL PHILOSOPHIES OF CULINARY HEALTH

In the excellent book *Cuisine and Empire*, Rachel Laudan details the history of many traditional culinary belief systems that developed after the advent of agriculture and investigates the underlying themes connecting them. Christianity, Buddhism, Judaism, and Islam would go on to radically transform these ancient systems all around the world, but not quite destroy them for good. Fragmented by the ages they have survived, pockets of ancient knowledge often passed down the generations by word of mouth, with little in the way of written doctrines.

Because of their mysterious and ancient origins, it is sometimes tempting to think of these systems as organic and benign. Whereas the dietary rules of religious belief are rarely adopted beyond those of faith, the ill-defined spirituality of Ayurveda or Chinese medicine is free to be assimilated into modern alternative culture, especially when it comes to beliefs about diet. These systems are largely based on principles of hierarchy, where everything on earth has its place on a well-defined culinary ladder. Plants feed from the land, animals eat raw flesh or vegetation, and only humans consume cooked grains and meats. To control what we eat is to separate ourselves from the animals. Cooking and agriculture made us civilized and led to trade and city states. In ancient religions around the world, cooking with fire is often linked to themes of purity, the heat of transformation signaling our distance from the savagery of the natural world.

There are common themes that underlie most ancient food beliefs, most notably the breakdown of the universe into a distinct and small number of essential elements. In the time of Hippocrates these were defined as air, fire, earth, and water, which related to the humors (blood, yellow bile, black bile, and phlegm) that circulate within the body. Whichever element or humor was dominant was thought to underpin disease, character, and numerous other qualities. Food consumption was thought to influence levels of the humors and so have the potential to cure or harm. Comparable themes run through the Ayurvedic systems from India and exist within traditional Chinese medicine.

AYURVEDA

Similar in many ways to the beliefs of ancient Greek medicine, Ayurveda centers on the existence of three *doshas* or life forces: *Vata*, *Pitta*, and *Kapha*. These are thought to pervade the cosmos and the human body, and it is their balance that defines the characteristics of every part of the universe. Vata is linked to breathing and air; Pitta, to fire and digestion; and Kapha, to water and health.

Related to the doshas are three temperaments, *Rajas*, *Sattva*, and *Tamas*, all influenced by the consumption of different foods. Rajas is thought of as active, hot, and passionate, contained within the blood and strongly present in most meats. Sattva is balanced, cool, and pure, present within the body as milk and semen and contained within rice, butter, and sugar. Tamas is inert and thought of as being heavy, stupid, dark, and evil. Its equivalent in the body is fat, and it is contained within stale foods.

It is widely believed within Ayurveda that a person's temperament can be changed by eating different foods, but it is also thought that the balance of the doshas locks you into a place in the cosmos. Not only was this used to enforce the societal hierarchy of high and humble cuisines, but it also led to the belief that the place where you were born is also the place where the balance of your humors is most in line with that of the universe. In the ancient world of Ayurveda, moving from your birthplace, either physically or socially, was thought to carry great danger of ill health. The poor worked the land on which they were born and ate the food of poverty because they wrongly believed that to do anything else would make them sick. It is easy to see how this system evolved to keep people in line, to grind down ambition and mobility, and to ensure the poor stayed poor and the elite were not challenged. It was an ancient wisdom that justified xenophobia and class division. It was also a culture of essentialism, where your diet defined your qualities as a human being, including every aspect of your health.

To transgress, to try to change your life, to accept anything other than the fate of your birth, was to fight against the will of the universe.

TRADITIONAL CHINESE MEDICINE

Traditional Chinese medicine is based along similar lines to the beliefs of ancient Greek medicine, breaking down the world into a small number of elemental qualities. Things here are slightly more complex, with a belief system based on five "phases": wood, fire, earth, metal, and water. Differing tastes, smells, and colors correspond to these different phases, as do different organs of the body. On top of this five-phase system, the medicine overlays the concept of balancing the opposing forces of yin and yang.

All disease is thought to be caused by a disharmony and imbalance between the phases and troubled interactions between the body and the environment. Diagnosis is complex and slightly impenetrable, depending on a number of specific interactions with a complexity that ensured the status and livelihood of practitioners. Often different treatments are advised for similar conditions, depending on the patient and the specific "pattern of disharmony." Most treatments are with herbal or animal preparations with the intention of restoring balance.

In this tradition, foods have the ability to maintain and restore balance in different ways. "Cooling" foods, such as eggs, tofu, and cucumber, nourish the yin and are used to treat fevers, anxiety, or vivid dreams. "Warming" foods, such as chicken, ginger, and vinegars, raise the yang and are used to treat lethargy, bloating, and stomach pain.

Why does anyone still use them?

Ayurvedic medicine and traditional Chinese medicine developed in a time when the functioning of the human body was poorly understood, and they are largely based on false assumptions and misunderstandings. The concepts of doshas, humors, or phases have no real-world basis and no genuine relation to human physiology. Ayurveda believes that toxins come out of your tongue and need to be scraped off. Chinese medicine still grinds up rhino horn to treat infertility and recommends mussels, prawns, and mustard for an upset stomach. The mechanisms and the majority of treatments are nothing but pseudoscience, and yet they are systems that have persisted for thousands of years and still exist today, despite the lack of evidence supporting them. In many cases, even when the supposed remedies have been shown to be ineffective or harmful, wellness gurus will regularly cite these wisdoms as fact.

The distinctly similar themes that apply across these systems, ideas of balance and harmony with your place in the universe, are clearly significant factors in their enduring charm. The concept of essential forces pervading all of existence is extremely powerful, an alluringly simple view of the world that appeals to our instinctive brains. But in many cases these ideas developed for the purposes of control and subjugation, to keep populations in check and prevent social change. They succeeded and thrived because of their powerful appeal, and although the reasons why they evolved are now largely irrelevant, this appeal has meant that they still survive in persistent fragments today.

The adoption by many new age and wellness movements over the years can also be explained by the lack of obvious connection to organized religion and the perceived absence of contamination by the corporate powers of the modern world. Science and evidence-based medicine are unfortunately associated with corruption, greed, and the pernicious influence of Big Pharma. For the political left and new age movements, this gives ancient belief systems vast appeal and a strange mystical quality, seemingly born before the dawn of time. It is perhaps ironic that the source of these beliefs, especially those of Ayurveda, lies within some of the most brutal social elitism imaginable.

Following ancient wisdom may make sense, but only if you don't trust the experts. Cognitive scientist and psychologist professor Steven Pinker told me:

> The antiquity of a bit of advice is a crude indicator that it has passed the test of time and may be better than a random bit of advice from some guy who just made it up. With the advent of science, we know that ancient advice is often malarkey, but that's because we have reason to believe that the methods of the scientists, and the self-policing of the scientific community, tend to align their beliefs with the truth more often than not. But in the absence of this verification, we reach out for any sign that someone knows what they're talking about, and antiquity is better than nothing.

In these circumstances, ancient wisdoms have great appeal, giving them the potential to override more sensible (but perhaps less fun) messages. There are other aspects of ancient wisdoms that might help them to create enduring belief, thinks food historian Rachel Laudan:

> The information that we have in the modern world on food and health is almost impossible for us to handle. Often it is not delivered in terms of the food we eat, it is more about complex micronutrients with strange names. We eat meals, not food pyramids, and the messages from the world of science can be hard to translate into meals. Chinese and Ayurvedic systems are more easily handled, more easily dealt with. When you have a cold there are a set of simple kitchen rules that you can apply. They make a certain

amount of sense, even today. They tend to encourage people to eat a little bit of everything and probably don't do much harm.

So, does it matter?

It is always interesting to consider whether holding a false belief actually matters and whether there is a potential for harm. For me, the main danger is that ancient wisdom is gateway pseudoscience, leading people down potentially dangerous paths. And although the food philosophies covered here are generally just messages of balance and variety dressed up to sound wise, there are many specific treatments in Chinese and Ayurvedic systems that are potentially harmful. Aristolochia, an herb used in traditional Chinese medicine, has been banned in many countries after being linked to kidney disease and cancer of the urinary system. In Ayurveda, Rasa Shastra, the traditional practice of adding metals to herbal preparations, is thought to be behind the often dangerous levels of lead and mercury contamination reported in many Ayurvedic remedies.

This is a book about food, not alternative medicine, so I will not discuss the subject in more detail, but there are many examples of people rejecting effective conventional treatments in favor of dangerous ancient pseudoscience. Chinese medicine has also had a damaging effect on a number of endangered species around the world, hunted to near extinction for imagined medicinal properties. It seems likely that accepting the dietary beliefs of ancient systems could easily lead people toward the more dangerous and harmful treatments so intimately connected with them.

WHAT CAN THEY TEACH US?

These belief systems have existed for thousands of years, so you cannot technically call them a fad. They did not develop to hide weight-loss goals, and although they do have a tendency to falsely medicalize foods, they do not do much dietary harm in general. What they do give us is an insight into the way that false beliefs about food can evolve and grow.

All ancient systems talk of mysterious essential forces that flow through everything—about balance, control, and essentialism. All see the food we eat as an outward expression of ourselves, rife with moralism and status signaling. All claim links to disease, giving a simple underlying cause to explain all medical suffering. All attach blame to ill health and tell the sick that their transgressions have brought their illnesses upon themselves.

The same can be said for modern diet cults. The most successful modern creations see a simple underlying cause for diseases. All look to blame the individual for their ill health, railing against the randomness of suffering. All are full of status signaling, perhaps longing for a new "high" cuisine, where green detox juices and alkaline water separate the privileged few from the feckless poor.

The concept that we are somehow out of balance is an appealing one. It defines a feeling that things are not quite right in some strange and undefinable way (perhaps a feeling that began with something other than food). We imagine that we could be better, feel stronger, be more alert and more vital. We can easily become convinced that we must be missing something, we must be slightly out of kilter with the world. Other people seem to have more energy, more vitality, and shinier hair, so we assume they must have discovered a hidden secret. Ancient wisdoms and modern health fads both offer to restore this mysterious lost balance—and the worried figure it's worth a shot. They're desperate to understand why their life is not quite as perfect as the media tells them it should be.

Beyond the link to modern pseudoscience, there is something important that enduring ancient wisdoms can teach us. As

Laudan points out, they do communicate their message in a way that appeals. They have a strong relevance to the world and the way people eat, linking food with health, and fitting into lifestyles with commonsense messages and instinctive appeal. Often they spread a message that encourages a balanced diet and a variety of different foods, rather than the restrictive fads of the modern day, and they give advice that fits into people's lives.

Scientific understanding is our best and most up-to-date wisdom, a symbol of the astounding progress we have made since these ancient systems were conceived. Yet despite everything it knows, all too often it fails to spread its message with the same clarity and relevance. We have nothing to learn from ancient wisdoms about dietary health, but maybe they can teach us a thing or two about PR.

Chapter Thirteen

PROCESSED FOODS

It is good to have something that everyone agrees on.

Food writer and journalist Michael Pollan: Don't eat anything your great-grandmother wouldn't recognize as food; if it came from a plant, eat it; if it was made in a plant, don't.

Tosca Reno, on the Dr. Oz website: Eating tons of foods that have been refined and heavily processed, often with added preservatives, coloring and other chemicals, can damage your health, zap your energy and pack on pounds.

Robert Lustig, professor of pediatrics and endocrinology, in a talk titled "Processed Food: An Experiment That Failed": In conclusion, the project has succeeded in getting the consumption of unhealthy food rising, and it has succeeded in cash flow for the companies, but it has failed dramatically when it comes to health. And there is only one answer: Real food.

Jamie Oliver: Knowing how to cook means you'll be able to turn all sorts of fresh ingredients into meals when they're in season, at their best, and cheapest! Cooking this way will always be cheaper than buying processed food, not to mention better for you.

Vani "The Food Babe" Hari: I cook often to ensure the quality of my food. I avoid fast food, refined sugar, factory-farmed meat, heavily processed food and questionable ingredients. I love superfoods like hemp seeds, chia seeds, and quinoa.

If there is one thing that every food writer, chef, and campaigner can agree on, it is that eating processed, factory-produced food products is wrong. To maintain maximum health, we must focus on eating real, whole foods and ditch the manufactured junk that has made us fat and sick. It does not matter if you are a clean-eating, new age health blogger, a celebrity chef, or a renowned and highly published researcher like Spector; the message is the same. For maximum health, we should be eating real food.

The popularity of this simple message is quite understandable. It fits well with our desire for easily followed rules and seems to make perfect sense. It is clear that we have become broken and sick because of our modern diets. It is quite obvious that through the march of progress we have lost touch with something pure and beautiful. Our relationship with food is broken, and to fix it we need to move back to a time of simplicity, when the food supply was straightforward, uncontaminated, and pure. We need to connect with nature, cook real ingredients, return to the ideals of a nuclear family, sit around the table joyfully embracing the sensory pleasures of a home-cooked meal, and enjoy a rich and happy life before the awful health crisis of the modern age.

Very few would argue—from naturopaths, clean eaters, Paleo freaks, and the low-carbohydrate, high-fat (LCHF) crowd, right through to dietitians, nutrition researchers, and medical doctors—that one unifying message for optimizing health through diet is clear: Ditch the unnatural, processed junk. Cook from scratch using local, seasonal, and real ingredients. Makes sense, right? But within this simple health message, there is a problem. A big fucking problem.

THE PROBLEM WITH RULES

And here is where I make myself unpopular. If my views on sugar or low-carb diets made you in any way uncomfortable, after this chapter you might want to plan a short rest. This is the part of the book where I criticize the doctrines of many of my great food heroes and, given my background, leave myself wide open to accusations of paid shillery. It is my belief that in demonizing processed and manufactured food products, not only are we falling for an oversimplified and flawed narrative, but we are in real danger of alienating the very people that need engaging the most. We are making exactly the same mistake that underlies every food fad we have discussed so far: attempting to classify foods as good and bad, and in doing so damaging our relationship with what we eat.

For the purpose of transparency, I have to admit vested interests at this point. I have spent over ten years working as a development chef in the food manufacturing industry, creating recipes for value-added convenience products for a number of well-known food brands. I love this work because I believe that the products I work on are important and a huge part of people's lives, capable of bringing moments of great joy. I also hold the thoroughly unfashionable view that when it comes to improving the dietary health of the nation, few people have more power than the food manufacturing industry to act as a force for good. Although in the world of food there are only a few outliers with similar opinions to mine, to some observers the strange obsession with all things natural, local, and unprocessed is curious. Steven Pinker, an informed mind on most subjects, says:

> Most of what I read about food is patently illogical, and seems to tap into a combination of status-signaling and primitive intuitions about purity, contamination and essentialism. Among other things, there is no nutritionally coherent category called "processed," and adhering to the locavore/food-miles religion would mean that I would never eat an orange again.

Within the extensive and sometimes brilliant work of Michael Pollan lies a strange irony. In his 2008 work, *In Defense of Food*, he rails against the concept of "nutritionism," arguing that many Western societies have become obsessed with the concept of dietary health, reducing food down to the effects of a few isolated nutrients and so damaging our relationship with it. In many ways I agree, as this reductionist approach doubtless leads to the sort of restrictive and rule-driven diets that I abhor. In the second half of the book, however, Pollan goes on to create a number of rules by which we should make our food decisions, including the two quoted at the beginning of this chapter. To underline his belief in rule-driven eating, in 2009 he published *Food Rules: An Eater's Manual*, literally a list of sixty-four rules to help us decide how we should eat.

There is much to admire in Pollan's work, not least his embracing of culinary joy and his desire for people to place pleasure at the center of food choices, but when it comes to the creation of a rule-driven approach to eating, I could not be further from him. Although simple, seemingly harmless statements like "avoid food products that contain ingredients that a third-grader cannot pronounce" and "avoid ingredients that no ordinary human would keep in the pantry" have great instinctive appeal, I believe they reveal deep underlying prejudices. Although the intention is to make the navigation of our food universe simpler, for many these rules will lead to guilt and shame about their choices.

Almost everyone who publicly talks about food and health spreads a version of Pollan's doctrine. Eat natural, eat seasonal, cook from scratch, eat real ingredients, avoid processed junk. The reason I pick on Michael Pollan is because, unlike some whom I may or may not have mentioned earlier, he is an intelligent and reasoned commentator, a sharp mind who doubtless weighs his words carefully. Yet even he can fall for the natural fallacy, a belief in a lost Eden where our great-grandmothers nourished the health of the next generation in perfect harmony with nature, the planet, and fresh, seasonal ingredients. And despite railing against "nutritionism," he has published a book of sixty-four rules telling us how to live our lives.

PROCESSED FOODS

Processed foods are broadly defined as any foods that have undergone a process to alter flavor, composition, or shelf life. So, processed food encompasses a wide variety of foods that even the most self-righteous health blogger would not suggest we avoid: peas, beans, lentils, quinoa, rice, flour, gluten-free flour, organic gluten-free flour, milk, yogurt, pasta, olive oil, virgin coconut oil, spices, dried herbs, chocolate, and couscous. They are all processed in some way. All these foods fall short of Pollan's measure, because all are made "in a plant." For all but the most crazed of fanatics, it is not sensible or practical to avoid every single processed food. We would have to spend our time pressing oils, grinding flours, and desperately drying and preserving ingredients so as to sustain a balanced diet. In doing this you would be processing these ingredients and so, presumably, if you are opposed to all processing of food, this would be unacceptable. Cutting is a process; cooking is a process. Heating, chilling, drying, and pickling are all processes. It could be argued that chewing is a process. If you have a fundamental opposition to the consumption of processed food, you need to find food that can be swallowed raw. I would suggest that water is your best option, and if you want to drink unprocessed water, then good luck to you, because water processing is perhaps one of the greatest life-saving innovations in the history of humanity (perhaps a useful reminder that not all processing is bad).

Ah, I hear people cry. This is just semantics! When people say to avoid processed food, they don't mean lentils and coconut oil, they mean processed junk. That's what people should be avoiding. This may be true, and I am not suggesting for a moment that your average health blogger is advocating the consumption of contaminated pond water, but the line must be drawn somewhere. We all eat processed foods, so it is surely just a question of which ones are allowed.

How about canned goods? Are they OK? Are they morally acceptable? Canned tomatoes are processed by heating (something known in the business as "cooking") to destroy microbes that would contaminate and spoil them. This extends their life and alters their flavor, but it is just heating, so surely that is fine. The preservation of tomatoes enables them to be stored and shipped around the world and consumed out of season by people who would otherwise not be able to enjoy them. They are packed full of micronutrients, contain only a low level of naturally occurring sugar, are fat-free, and, more important, are completely delicious. For some uses, the flavor of canned tomatoes is superior to fresh. Surely canned tomatoes pass the test, something that campaigners opposed to processed and manufactured goods would approve of. Something very much made in a plant that we can happily, guilt-lessly consume. But what if I told you that to preserve the life of most canned tomatoes, manufacturers add something called an acidity regulator, namely 2-hydroxypropane-1,2,3-tricarboxylic acid, a chemical additive that most third graders would struggle to pronounce? Is that still OK? Is it OK for manufacturers to take a perfectly natural food, add chemicals to it, risk toxic contamination by placing it into a can, and process it to such an extent that its shelf life will be eighteen to twenty-four months? When looked at this way, maybe canned tomatoes are a vile, needless Franken-food with no place in our diets.

Maybe the real argument is about "ultraprocessed foods"? Are those the ones we should be avoiding? Ultraprocessed foods are loosely defined as any packaged foods comprised of several ingredients, including substances not generally used in cooking. Again, the definition is quite loose and will most likely encompass a number of products of poor nutritional value. But is it really sensible to classify all foods in this category as unacceptable? What are we really saying in making that broad classification? That a jar of pasta sauce is harmful to our health? That we should never eat a chocolate bar or a bag of potato chips because they are somehow "unnatural"? That we should never eat in a restaurant? That we should never buy a loaf of bread?

Throughout any day we will consume thousands of different chemicals in the form of food. All food is composed of chemicals, and most of these chemicals will be unpronounceable by a third grader. Just because that combination of chemicals comes from a natural source does not imbue it with some sort of magical health-giving powers or ensure that it is completely safe. When they are swirling around in our digestive system, our body has no way of telling if the 2-hydroxypropane-1,2,3-tricarboxylic acid molecules were added in a factory as an acidifier to extend the shelf life of a pasta sauce, or came from a squeeze of lemon juice, where this chemical is known by its common name of citric acid.

The health-giving properties of the food we eat are determined by their chemical composition, not by some magical origin story. There is no fairy dust of naturalness that makes home-cooked (or maybe "home-processed") meals healthier than anything made in a factory. At their most basic definition, fruits, vegetables, and meats are complex combinations of different chemical substances. That combination is the result of the random flow of evolution, not some magical design of nature. Nature has no wisdom. It is random, strange, and undirected. The vast majority of poisons we are likely to encounter, including the most damaging ones of all, are made by natural processes. We may be morally troubled by food made in a factory, but the nutritional value and health-giving properties are not defined by the type of building it was made in or by the size of the company that conceived it. In creating arbitrary rules around food, based on the story of its creation, we are moralizing about food choices. The moment we do that, we risk creating shame around food, alienating our neighbors, and damaging our bodies—the very thing we've been trying to avoid.

The truth is that our food supply is safer than it has ever been. There is less contamination, and there are fewer cases of poisoning or health problems caused by diet than at any point in history. If we dropped in on our great-grandmothers at a typical mealtime, we would find people scraping an existence with nutritionally poor, unbalanced meals and food scares that make horsemeat contamination look like a pajama party. Most of our great-grandmothers probably lived much of their lives in the first half of the twentieth century. Diets at this time in the US and the UK were largely starch- and meat-based, with over 50 percent of calories coming from bread. There was little understanding of the concept of vitamins, meaning that deficiency diseases were rife. The lack of safety checks and awareness meant that gastrointestinal disorders were also far more prevalent than today and far more likely to have serious consequences. There was no cold chain for the storage and delivery of foods, meaning that fresh fruits, vegetables, milk, and eggs were scarce, and at certain points in the year virtually nonexistent. Our great-grandmothers could expect to live around two thirds as many years as we do, largely because we enjoy an increased awareness of dietary health, increased safety and integrity in our food supply, and industrial processes that have made our lives easier, healthier, and richer. By pretty much any measure you can think of, the golden age is now, and yet we remain convinced that we are broken.

Why is this the case? Much as with the denigration of the present seen when talking about toxins or Paleo, it is clear that we have an inclination to idolize the past. As we age, we confuse our own biological decline with that of the world. We are also highly inclined to believe in the wisdom of nature. After all, if nature can create something as wonderful and miraculous as the human brain, perhaps the most complex structure in the universe, way beyond the abilities and comprehension of even the most skilled human engineers, then surely it must be wise and better able than us to define what is healthy.

Sydney Scott is a researcher working with Paul Rozin at the University of Pennsylvania who looks at the preference for natural products. "Many consumers think nature is good, sacred, benign, and gentle," she says. "Especially as nature is perceived as safe. Some of these beliefs might have a religious basis; nature is God's work and therefore sacred, good, and our duty to protect. Some beliefs might have to do with perceiving nature as 'familiar'— something that humans have interacted with for centuries—and therefore inherently more 'known' and 'safe.'"

The concept that nature is benign and sacred harks back to the romantic idealization of the natural world, a belief in its purity and goodness. It is the reason why Charles Darwin and Alfred Russel Wallace's work on natural selection was so shattering and divisive; it revealed nature's true colors, full of pain, hunger, and desperation in the battle for survival. It also explains why, when German chemist Friedrich Wöhler first synthesized urea in 1828, his seemingly simple experiment challenged our understanding of the natural world, perhaps forever damaging our inclination to religious belief. Until then, living organisms were thought to be imbued with a natural "vitalism," a mystical property that distinguished them from man-made chemicals. Wöhler managed to synthesize urea, a "natural" substance, from an inorganic salt, proving that all living things were in essence little more than a bag of chemicals, reacting, changing, and excreting waste. Wöhler's work promised a future where all the reactions of living things could be replicated in a test tube, opening up the new possibility that the only mystical thing about life is its huge chemical complexity. The distinction between *natural* and *chemical* was revealed as nothing more than a conceptual one, formed in our minds to help mask the mechanical and seemingly purposeless nature of life.

Despite our advances since then, that distinction still lives on, and so does our idolization of the natural world. Even for those without religious belief, there is still something of the divine about it. It seems designed and perfect, but in reality it exists due to a series of random genetic mutations. These mutations may produce a useful change in the way our body works, or they may cause cancer—we only witness the survivors of this process.

We imagine natural foods as "pure," and yet they are not pure at all. The only things that we consume in a pure form are water, sugar, and salt, and these are purified by our hands. Nature is impure, imperfect, and glorious for it. We grew to exist among it and, since the dawn of humanity, our ability to flourish depended upon us manipulating it for our nutritional needs. Scott believes that "the grass is always greener on the other side, so when we are separated from nature, as many Western societies today are, we view it as an ideal. When we are grappling with nature every day, as we were in centuries past, we view technological advancement as an ideal."

Scott's work also looks at the nature of sacred or protected values.[1] These are values held that are not affected by evidence, usually based on disgust and pertaining to sex, food, and the violation of the body. We fear ingesting something seen as unnatural and causing contamination; for many it violates our most deep-seated values. We will demonize perfectly healthy foods based on these unjustified fears. We believe that a factory will somehow contaminate food with a sense of the unnatural, that it violates the wisdom of nature, that it destroys some of the goodness contained within. We fear progress, fear chemicals, and attach emotive terms like "Frankenfood" or "fake foods" to spread our sacred values to the world.

THE STRANGE ALLIANCE

Most of all we fear progress. We fear the march of time, and as it ravages our bodies and minds we mistake our decline with a decline in the world. We long for imagined times past when all was pure. The disaster of modernity is seen as an inevitable fact by all observers, and yet it bears little relationship to reality. We refuse to accept that the golden age is now, that this is as good as things have ever been. It is not perfect, but every society that has ever existed would eagerly swap their lives with someone living in the developed world today. We are safer from hunger, disease, war, and crime than we have ever been.

On the left of the political spectrum, the vilification of processed and manufactured goods is seen as a problem of corporate corruption, the poisoning of the world driven by an insatiable desire for profit and growth. The manufacturers are painted as a force of unspeakable evil, making us fat and sick, lining their greedy pockets at our expense. The left longs for a time past when small businesses and farms met our food demands, simple cottage industries crafting beautiful goods with love, care, and personal attention.

The political right rail against processed goods, too. For many on this side, the processing of foods is against the work of God, against the beauty and purity of nature, messing with things that we do not understand. For others it is driven by a simple desire to return to the good old days, the time of our forebears, a time of purity and conservative values, before selfie sticks, low-cut jeans, and smartphones ruined the world. A time when men were men and women knew that their place was in the kitchen. Before convenience and modernity freed women to live fuller, richer lives.

Little has done more than prepackaged meals to liberate women from servitude, freeing them from the time and cognitive effort required to create an endless stream of home-cooked meals from scratch. Processed convenience food has set women free, and every time we criticize convenience choices, we are showing our desire to drag women's bodies and minds away from the workplace and back into the kitchen. We reveal that this is where we think women belong, and we label those who choose otherwise as slovenly, selfish, and neglectful of their duty. Nourishing family meals cooked by a loving mother and eaten together around the table have become a symbol of all that is good, and their decline the reason for all of society's ills. Those who do not achieve this goal are seen as pariahs, yet for many these demands are unrealistic.

The left and the right of the political spectrum create an unlikely alliance in their attitude to convenience food, and that alliance leads reasonable people to confirm their belief in its truth. If such disparate commentators agree, then surely it must be true. Not one campaigner for healthy eating dares to embrace convenience foods, as to do this has drifted so far from the consensus as to be unacceptable. Yet to campaign for healthy eating based on the assumption that the only option is a lifetime of home cooking is profoundly unrealistic. It does not meet people where they are, it sets unreasonable goals and expectations, and it drives unhelpful associations. It attaches guilt and shame to perfectly reasonable food choices.

FUTILE ADVICE

If guidelines and advice about healthy eating are too far away from the way people live, they are more likely to ignore them completely. For this reason, making such unrealistic expectations as outlawing convenience foods and insisting everything is made from scratch using seasonal, local produce is a foolish and point-less activity. People are living busy, stressful lives. You may as well be asking them to jump to the moon.

Tell a young mother living a shitty life in a shitty housing project, surrounded by poverty, drugs, and crime, that if she just cooks everything from scratch, then she and her child will be really healthy—she will look at you like you just landed from Mars. The best response you can expect is "I don't fucking care; my life is shit anyway." There are communities, towns, and cities where whole populations are constantly labeled as feckless, bro-ken, and beyond redemption. Try walking into a community like that and tell the residents that organic meat is affordable if you cut down on some of your treats and food waste. If the world keeps on telling you that your life is shit, eventually you start to believe it. Caring that a poor diet might be affecting your long-term health is fairly low down on the list of stuff to worry about. When you are not sure how you are going to make it to the end of the week, worrying about your saturated fat and fiber intake doesn't really make the list.

Every time we demonize the processed foods that people eat, every time we label essential manufactured products as dirty or unclean, every time we tell people that their choices are shit and they are killing themselves and their children, we are widening the divide. We are trying to impose aspirational middle-class food values on people who don't want them, and in doing so we are marginalizing those who need help the most. And we do this for no good reason. These widely held views on processed foods are based on pseudoscience and generalizations as ill-conceived as the alkaline diet or the detoxing properties of cucumbers.

It pains me a great deal to admit this, but many people are just not that interested in food. To cook from raw ingredients may be simple and effortless for a chef or food obsessive, but for many it is stressful, joyless, and difficult. It is not just about time. Although some meals can be made quickly—and the repertoire of a celebrity chef is not complete until they have made a book titled *Real Fast Food* or similar—these ignore the sheer cognitive effort required by many people to put food on the table.

After a hard and stressful day, we may still have the time available to cook, but sometimes putting a pizza in the oven is just easier. This is not stupid or illogical, and it is certainly not immoral. Perhaps it allows us to spend time and effort on something more enjoyable or important. There is (for some) more to life than the food we eat. There are different things to concentrate on, different journeys to make that are equally nourishing for the soul. There are books to read, movies to watch, games to play, and races to run. To believe that we will always have the time to cook is to misunderstand what it is to be alive in the modern world.

As I write these words I am sitting in McDonald's on a bright summer's evening toward the end of the school holidays. I am watching a succession of families returning from action-packed days, many irritable and fractious, yet all full of irreplaceable and joyous memories of time well spent. They are rounding off their day with the unhealthiest of fast foods: burgers, fries, and sugar-laden drinks. They are doing this for pleasure and because they want to be fueled without having to think about it. They are willing to trade away the healthier option of a home-cooked meal because the memories they have created today are vastly more important. This is not an illogical choice; it is more than worth the trade. Convenience food enhances lives because it frees people to live them how they choose.

The key to improving the quality and healthfulness of the food people eat lies in an engagement with food manufacturers, not a rejection of them. Companies that deliver processed food have great power to provide options to improve dietary health, and yet increasingly those companies are being ignored and sidelined. When public health bodies engage with the food industry, they are vilified, too, told that they are colluding with a corrupt and malevolent enemy. Food manufacturers and retailers have great power—far more than any celebrity chef—to help and improve people's diets. They can offer sensible, realistic solutions that fit into modern lives. Although the food industry should be held accountable for transgressions, if just once a chef or campaigner praised them when they did something positive, it would make my heart sing. I long for a time when campaigners and manu-facturers present a united front. Imagine the power and force for change that could be created, offering and endorsing sensible options to improve people's lives.

Convenience foods are already with us, thoroughly integrated into our lives, invigorating them, enlivening them, and allowing us to live them to the full. To reject them and the modernity they represent is completely unrealistic. To attach guilt and shame to them, to ascribe moral values to those who choose them, is a dangerous path. At best it will push people toward negative behaviors, as guilt and shame have been shown to do. At worst it will damage people's relationship with food for good. This brings us to **Rule Number 9 in the Angry Chef's Guide to Spotting Bullshit in the World of Food**: *They will falsely idolize the time of their grandparents and great-grandparents.* They will tell you that things were better back then.

Chapter Fourteen

CLEAN EATING

A GUIDE TO FIGHTING THE UNSTOPPABLE MULTIHEADED HEALTH HYDRA

And so we come to clean eating, the chapter that in many ways is the heart of this book. It is where my story started, sitting listening to the illogical ramblings of one of clean eating's high priestesses at a food industry conference. Until I encountered her, despite being someone who prides himself on knowing a bit about food trends, I was completely unaware of this new culinary beast, poised though it was to catapult into the mainstream. For clean eating is a new type of creature, distinct from the clear origins and written manifestos of the diet trends we have discussed so far. "A lifestyle, not a diet," we are told, proud of its holistic approach, lack of hard rules, and selective embracing of occasional fragments of science that suit its needs. It is the health trend of the modern era, ill-defined yet hugely influential, the very expression of new media's undirected and uncontrollable power.

For sensible voices such as medical doctors, dietitians, and pseudoscience debunkers, this makes engaging with clean eating as hard as nailing gluten-free jelly to a wall. It is little more than a loose conglomeration of social media savvy and self-appointed gurus, each with a different interpretation, each with a different doctrine, and all hiding diets of restriction behind veils of holistic wellness. It is an ever-growing multiheaded monster without an easily located heart, seemingly impossible to fight, and yet so strongly entwined within our modern media that it is in danger of slowly strangling out sensible voices forever.

It is hard to fight, but perhaps not impossible. It does have some weaknesses, and it does leave clues that can help us reveal its true nature to the world. But it is a tough beast to slay, especially because of its sweet and innocent face. The naysayers are often damned by those who believe that the beast has been sent to save us from ourselves. Many ostensibly intelligent commentators seem unable to see that behind the squeaky clean mask of kale and false promises lies a big, dirty secret. As with the criticism of many alternative practices, criticism of clean eating is perceived as a restriction of people's free choice, and that sort of restriction is always seen as bad.

THE SOURCE OF THE OUTBREAK

First, let us try to look for the origins of this curious trend. Although many will look back to the whole foods movements of the 1960s and 1970s, I think the real origins of eating clean, perhaps ironically, lie in the efforts of the food manufacturing industry to "clean up" its product formulations in the 1990s. Convinced that consumers were becoming suspicious of the many unknown chemicals listed on food packaging, retailers and food brands became obsessed with creating "clean label" products, taking out many additives that consumers deemed unnecessary. Although holding food manufacturers accountable for dishonest practices is clearly a good thing, this desire for a clean label often ended in the selling of false promises. Harmless and useful preservatives, emulsifiers, flavor enhancers, and even on occasion vitamins were removed so as to create seemingly more wholesome products, replacing them with "natural" alternatives. Often these replacements only benefited consumers by appearing less intimidating. Monosodium glutamate (MSG) might be replaced by a yeast extract or a natural flavoring with similar properties caused by identical chemical compounds. Chemically modified starches were replaced with heat-modified ones, because the latter can be labeled as "natural," despite being of nearly identical composition. This did not happen because the replacements gave any health benefits, but because they made for a better-looking, cleaner ingredients list. "Clean" became the marketing buzzword underlying many a product launch and reformulation, despite being little more than a slightly disingenuous rearranging and redefinition of constituents. When I first entered the food manufacturing industry as a development chef in 2005, this drive constituted a huge part of the workload, finding ways to replace items that did not look as clean on a label, and only using ingredients that could be described as pantry items on an ingredients list. Never once did I actually feel that I was making a product healthier as a result of these changes.

We shall return to chemophobia and the fallacy of naturalness later on, but the result of all this was that it helped to create a belief that the modern food supply *needed* to be "cleaned up" in some way. This would come back to bite the food manufacturers on their clean-labeled backsides a few years later.

TOSCA RENO

In 2006, a Canadian fitness model named Tosca Reno published a book called *The Eat-Clean Diet*, taking the language that food manufacturers had created and using it to brand a fairly unremarkable diet and exercise plan. What resonated and helped make the book successful, though, was Reno's assertion that her plan was not a diet in the normal sense: it was more of a movement, a simple, seemingly logical blueprint for changing the way you eat. It embraced the purity of "clean," unmanufactured produce, rejecting the modernity of processed foods completely. Despite actually being a fairly standard diet book, Reno's genius was to realize that the way dieting and weight-loss writers had made a living for so long was distinctly outdated. To the shiny new millennials, the boredom, banality, and restrictive nature of a diet was now considered distinctly low rent. When you have grown up being able to access a world of information through a tiny handheld device, when you can translate any language at the touch of a button and video call anywhere in the world instantly, it is not surprising that you might come to think that health and well-being should be something to be achieved effortlessly. Tosca Reno's "eat clean" diet promised just that.

It is understandable that many people might have found this simple message quite compelling. It took the "clean" promises of the food manufacturers to the extreme, advocating a diet based completely on unprocessed foods, cutting out anything that has a label, anything manufactured.

In some respects, dietitians and people working within nutritional science might actually find some common ground with the principles that underlie Reno's plan. Many processed foods are cheap, easily consumed sources of calories, low in micronutrients and fiber. Additionally, cutting down on processed foods might make people generally inclined to become a bit healthier. Reno advocated the eating of lots of whole foods and vegetables, steering clear of the food group exclusion that was to pervade clean eating as things progressed. Most of the recipes she talks about are reasonably balanced, and much of her advice could even be described as sensible (if a little unrealistic for most of us). But even in clean eating's earliest days, there was some stuff that did not add up. Reno talks about how calories don't matter (they do), about the need to detox, and about the acid/alkaline properties of foods. Much of her blueprint is based on what is called the naturalistic fallacy, a desire to reject modernity without a clear explanation or evidence. This was the start of the narrative that manufactured foods are making us sick and the only thing that we need to be naturally healthy is to reject them completely. The seeds were sown by her book, and these seeds were to flourish into something far bigger.

THE KEYS TO SUCCESS

I do not think that it was Tosca Reno's intention to create a powerful movement—I think she just wanted to sell a few books—but for a number of reasons, the way her diet was marketed to the world created a compelling proposition, destined to evolve and grow.

First, it was never sold as a diet plan in the conventional sense, instead taking the form of a movement and belief system. As such, anyone was free to interpret it however they wished. In an age where there are so few barriers to sharing information with the world, that loose doctrine and capacity for self-interpretation had great appeal. Although this has always had the potential to happen, social media has created an environment where like-minded people can easily find each other, meaning that when Tosca launched her book the potential for a movement to proliferate quickly was greater than ever. Clean eating was very much a movement of its time.

Second, eating clean has great appeal to the instinctive brain, creating a simple narrative of good versus bad. There is no need to count calories, no need for monitoring or control. Simply follow a couple of basic, easily remembered rules, and you will be healthy.

Third, and perhaps most worryingly for me, is that it probably works fairly well.

What? This chapter has taken an unexpected turn.

If short-term weight loss is your goal, then Tosca Reno's diet, which was after all initially sold as a method for losing excess fat, is probably not the worst diet in the world. Pointless pseudoscience aside, for most people, her plan may well result in losing a few pounds. Reno was quite clear about the weight-loss goals at the heart of her book, and it carries the distinctly old-school alliterative subtitle "Fast Fat Loss that Lasts Forever." Her assertion that calories do not matter is false, but what is probably likely to work when it comes to weight loss is the creation of eating rules that might just restrict followers from a number of calorie-dense items that they usually include in their diet. People will tend to eat less and lose weight when given such rules to follow, especially when you affix powerful, emotive terms like clean (and by association dirty) to the food they eat.

These terms are another key to the diet's effectiveness. As with many brands, a huge factor in clean eating's success has been in the appeal and power of its name. Food is neatly separated into clean and dirty, and these terms have the potential to strongly affect our emotions and behavior. As Steven Pinker discusses in *The Better Angels of Our Nature*:

> I have mentioned . . . the mind's tendency to moralize the disgust–purity continuum. The equation holds at both ends of the scale: at one pole, we equate immorality with filth, carnality, hedonism, and dissoluteness; at the other, we equate virtue with purity, chastity, asceticism, and temperance. This cross talk affects our emotions about food . . . Since the human mind is prone to essentialism, we are apt to take the cliché "you are what you eat" a bit too literally.

ARE YOU WHAT YOU EAT?

The phrase "You are what you eat" is a commonly held wisdom

within the bullshit-nutrition community. Of course we are not what we eat. Having evolved as omnivores, our bodies have a remarkable adaptability to a variety of different diets, and we can consume a wide spectrum of different foods with very little impact on our body composition. Vegans are quite clearly made of meat and contain a remarkably similar mix of proteins, carbohydrates, and fats to the rest of us.

The phrase originates from the French gastronome Jean Anthelme Brillat-Savarin, yet over the years it has been paraphrased and twisted from its original meaning. Brillat-Savarin originally said, "Tell me what you eat and I will tell you what you are," and most likely never intended the literal connotations ascribed to it by modern health gurus. He was really referring to how the food we eat reflects our character and how our culinary choices project our values to the world. Strangely perhaps, given it has become the mantra of the food-as-health community, this insight underlies some of the reasons for the proliferation of misunderstandings and lies in the world of food.

When it comes to clean eating, creating an association with cleanliness and purity in food is likely to imbue the eater with feelings of a similar nature. To break the rules is to eat something dirty and to become dirty yourself, creating feelings of disgust. It is perhaps this powerful association that has given the clean-eating movement its greatest power to evolve and grow. Tapping into these feelings and creating a strict morality around food choices gives followers the impetus to stick to the rules, to follow the diet, and to feel virtue and superiority over those who do not.

And this I feel is clean eating's Achilles' heel. Followers are signaling their purity, their morality, and their goodness. By implication, those who do not follow are the opposite. The word clean is the key to the movement's power, yet it is an outward signal of its true pernicious nature. For clean eating has evolved to sell lies and justifications for disordered eating patterns. It has created a moralizing culture of restriction and shame.

THE BIG, DIRTY SECRET

As clean eating has grown in power, the weight-loss goals that lie at its heart have all but disappeared from show. But make no mistake, it is aspirations of thinness, not wellness, that drive its success. The new breed of clean-eating bloggers talk about a "lifestyle, not a diet" and rarely if ever mention weight loss, but that is what is underlying an outward concern for our holistic well-being. Clean eating hides a desire for thinness and sculpted, self-conscious beauty, all to be achieved seemingly without effort. The young women who have come to encapsulate this trend parade an endless stream of pictures on their Instagram accounts, signaling their purity, their morality, and their enviable ability for control. But most of all they show their thinness. All of them will post as many pictures of themselves as they do of their food, carefully crafted Instagram shots designed to show us their ability to control their bodies, their ability to be effortlessly skinny. It is a distinctly retrograde world where women signal their glamour, health, and domesticity and strive to display the moral purity of their lives.

The problem is, the sort of aspirational thinness they present is hard to achieve. To follow the simple rules of Tosca Reno's original clean-eating plan would be unlikely to accomplish the sort of fast, dramatic results demanded of an image- and weight-obsessed society. And so, what sells itself as a happy, healthy lifestyle has evolved into an old-fashioned diet of restriction. The new stars of clean eating advocate that followers cut out whole food groups, forcing the same powerful dirty/clean associations onto any number of perfectly healthy foods.

Gluten is rejected by most, outwardly because of pseudoscientific assertions that it damages the gut of nonceliacs, but in reality because it forces followers to cut out many staple foods. Many call for a rejection of grains, potatoes, and other sources of carbohydrates, stealing from the likes of Paleo with vague antimodernistic notions of our bodies not being able to tolerate modern industrially produced foods. Some categorize all dairy products as unclean, borrowing from the alkaline diet, claiming that it causes acidity and draws calcium from the bones. Some reject meat, fish, and other sources of protein for similar alkaline-based reasons. Most advocate detox in some way, forcing followers into highly restricted fasts based on false notions of toxicity. What is there left to eat, when you are not allowed bread, potatoes, or grains; fish, meat, or any dairy? The most important part of anyone's diet is variety, which is the exact opposite of this narrow prescription, devoid even of most sources of protein and carbohydrate. Of course, these dietary restrictions are sure to result in weight loss, because a diet where whole food groups are removed is very likely to result in lower calorie intake. The power of clean eating comes from imbuing forbidden foods with a sense of dirt and impurity, resulting in them being rejected all the more vociferously.

And this is why clean eating is such a repugnant and potentially damaging trend. When people try to hide the weight-loss goals that lie at its heart, they need to justify the arbitrary rejection of certain foods by other means, and this is where the pseudoscience comes in. With sweeping certainty, the principles of various bullshit-peddling diets are incorporated to justify the need for control. Clean eating is the pseudoscience magpie, stealing items from all corners to create a vague doctrine. But underlying it all is restriction and control, powered by notions of impurity and disgust. For someone who has spent his life immersed in the world of food, celebrating everything that it has to offer, a movement that demonizes life's greatest and simplest pleasure is an affront to everything that I stand for.

I spoke to Judy Swift, associate professor of behavioral nutrition at Nottingham University, who is not short of opinions on such things. On the power of these arbitrary rules, she told me,

> The nature of controlling and creating rules gives us a sense of civilization. Food rules separate us from the animals, leading us to often use food symbolically. We are told that we can manipulate food, that we are not constrained by the environment. This is why food rules are so often used by religions to prove spirituality. Clean eating is the same, it is about proving how morally tough and self-controlled you are.

THE CULT OF THE OBSCURE

Perhaps to cover its restrictive heart, clean eating is also defined by its hunt for the obscure. It embraces many unusual, expensive ingredients, eulogizing with great certainty about the health-giving properties of many a poorly understood micronutrient contained within chia seeds, quinoa, spirulina, miso, bone broth, goji berries, coconut water, wheatgrass, or baobab. Driven by misunderstandings of science (particularly concerning antioxidants), these ingredients are imbued with magical properties, the cleanest of the clean. The more obscure, expensive, and harder to obtain, the better. Organic is considered essential by most, others insisting on biodynamic.* For those who include meat, it must be grass-fed. Fish must be wild. Milk should be sourced raw, lest the pasteurization process destroy its naturalness.

Biodynamic agriculture is an early form of organic farming developed in the 1920s by Rudolf Steiner. Although many of its practices are similar to more conventional organic methods, it encompasses a number of mystical and spiritual elements. Among its stranger beliefs is the use of an astrological lunar-planting calendar and the burying of quartz-stuffed cow horns or the yarrow-stuffed bladders of red deer so as to harvest cosmic forces in the soil.

Clean eating seems deliberately designed to be expensive, exclusive, and difficult to achieve. Judy Swift explains,

> Much like with any fashion, these things are driven by status signaling. They are saying "I have the resources to do this," both financial and cognitive, keen to give off signals, driven by an insecurity about their social position. Being seen to know about and consume an obscure ingredient gives off a certain message.

Similarly, Britt Marie Hermes, our reformed naturopath, told me that within the world she used to inhabit, one of the great appeals was that "you feel privileged, like you hold a secret knowledge, and that creates delusions of grandeur. It is a distorted reality."

This embracing of obscurity helps to create a feeling of community within the world of clean eating. Followers are shown a new path, with special information being imparted to them. Those who do not follow the path are seen as unclean. There are worrying parallels with religious bigotry. For followers of clean eating, theirs is a secret world, and the internet gurus and celebrity followers form the priesthood of their religion. Although their belief requires them to reject conventional science, to disregard advice from the medical doctors, dietitians, health bodies, and governments, they are happy to do this. I spoke to Steven Pinker about the conditions required for people to make this seemingly illogical step. He told me:

> One is people's tribal affiliation. If they think that the "authorities" are just a self-contained, self-serving priesthood, or worse, a conspiracy of capitalist profiteers, rather than a disinterested source of objective truths, then they'll

dismiss their advice and go with the advice of a priesthood they're most sympathetic to.

Clean eating is a tribe in the most modern sense of the word, a loose collaboration of people connected by internet affiliations, modern communication methods, and a shared set of beliefs. A tribe that is conditioned to accept dangerous pseudoscience to help spread its message of dietary restriction. A tribe that is happy to use the language of eating disorders to sell a false promise of effortless thinness. In a world battling with an epidemic of obesity, clean eating's association with health provides it with a veil of responsibility, yet its message can be one of great harm. When Kate Moss claimed that "nothing tastes as good as skinny feels," it was clear how harmful this statement was, but at least it came with a level of honest self-awareness. Clean eating, with its arbitrary restrictions, moralistic language, and pretense of holistic wellness, is just as or even more harmful. Yet, it's more likely to sneak into people's lives, to overtake and control those genuinely wanting to improve their health. As we shall learn, for those vulnerable to obsession and disorder, clean eating can be a vicious and overwhelming beast.

Chapter Fifteen

EATING DISORDERS

As I sat in the bleak, windowless therapy room and pathetically mouthed happy birthday to Becky, I wondered if this is how she always imagined she would spend her twenty-first birthday. She seemed happy enough, but then again, dinner time was a few hours away yet. How did I end up here? All I did was decide to skip breakfast and Anna Wintour does that every frickin' day—she doesn't have to have someone watch her take a dump. Surely I'm not as bad as the boney women who sit in a circle of misery around me, the string from their paper party hats digging into their sinewy necks. I didn't eat my beans one at a time like Becks did, I didn't hide pieces of toast like Mona and I never left a vomit stench in the toilet after every meal (like some). A singular cupcake was brought out tentatively in celebration of Becky's birthday. There was no pressure to eat it, but if we wanted to, we could. I still couldn't.

—Eve Simmons, eating disorder sufferer

I want to make one thing quite clear: I do not believe that clean eating, or fad diets of any sort, are the cause of eating disorders. Although there may not be a causal connection, there is, however, a clear link between the two. In a 2015 newspaper interview, Dr. Mark Berelowitz, an eating disorder specialist at the Royal Free Hospital in north London, said that between 80 and 90 percent of his patients were following clean-eating diets, excluding sugar, meat, dairy products, carbohydrates, or gluten. Anyone working with eating disorder patients will (anecdotally) tell the same story. The problem with the rules, restrictions, and moralizing language of clean eating is that for anyone whose patterns of eating are inclined to disorder, these things have a huge attraction. Like many sufferers, Eve Simmons (quoted above) believes that an obsession with clean-eating websites drove the onset of her disorder, a swift and terrifying descent into life-threatening illness.

But as I force myself to repeat every day, correlation does not always imply causation, and without definitive evidence, campaigners should resist blaming clean eating as the cause of eating disorders. The reality is far more complex and nuanced, and the simplistic tabloid image of obsessive teenage girls driven to disorder by images of skinny models is a long way from the whole truth.

This does not mean that I am about to let anyone off the hook. Far from it. But in understanding how clean eating interacts with the complex psychological afflictions of eating disorders, it is important to understand a little bit about the disorder itself.

A HIDDEN EVIL

In a world preoccupied by an epidemic of obesity, the horrific cost of eating disorders is often overlooked. It is, however, a significant and serious problem, and it is to our shame that it is sometimes buried under a deluge of other food-related health issues. The monetary cost to the economy is difficult to tally, but it isn't hard to imagine the toll of lost income combined with the cost of treatment, which averages $30,000 per person. In the UK, it is estimated that eating disorders cost the economy around £15 billion per annum, according to a report by auditors at Pricewaterhouse-Coopers.

It is estimated that 24 million people are affected by an eating disorder in the US and millions more worldwide. Although it is not an exclusively female problem, only 11 percent of sufferers are male. Of all sufferers, 10 percent are anorexic, 40 percent suffer from bulimia and binge-eating disorder, and the remainder are not specified.

Anorexia is described as a condition where someone tries to keep their weight as low as possible by extreme restriction of food intake or exercising excessively. Bulimia occurs when someone goes through periods of binge eating followed by self-induced vomiting or laxative use in an attempt to control their weight. Binge-eating disorder is when someone feels frequently com-pelled to eat excessive amounts of foods in a short space of time. Within these groups a new disorder has been described, that of orthorexia, where a sufferer will become fixated with stringent rules, cutting out many items perceived as unhealthy, and engag-ing in a damagingly restrictive diet.

The average amount of time that the brutal disorder of anorexia affects sufferers is eight years, and it often lasts much longer. Although 46 percent of patients will fully recover, a staggering 20 percent will die, the highest mortality rate of any mental illness; 40 percent of these deaths are from suicide, the rest from the inev-itable organ failure caused by lack of adequate nutrition.

Behind the statistics lie many thousands of desperate and tragic stories. Eating disorders are a descent into the abyss, a grim and consuming illness, with its victims slowly destroying them-selves in the sight of those who love them the most. This descent tears at the most primitive desires of those who care, robbing them of the ability to provide adequate sustenance and health. Unlike many diseases of childhood, they are the cause of much public shame and admonishment, with mothers and fathers left feeling that they have failed in the most basic of parental tasks, and igno-rant observers speaking of silly little girls who need to pull them-selves together. As long as these conditions are misunderstood as a teenage vanity, more suffering will be inflicted on innocent families at a time of great need. I define myself by my love of food, and seeing people's relationship with what they eat become so broken sends a chill to my very core. When the grip of the illness is so tight that sufferers are willing to starve and deny their own body, despite access to a world of culinary joy, it seems to me the very embodiment of tragedy. I am sad when plates of food I have cooked are not licked clean, so to feel that my ability to tempt and delight with culinary creations would not be able to reach someone gnaws away at my soul. Eating disorders are shattering, merciless conditions that destroy the lives of many bright, beautiful, yet deeply troubled souls. Anyone who has been touched by them will know that they are devastating, each case capable of breaking apart the strongest families, eroding the will of the most valiant, and cruelly ending countless young lives.

THE DRIVERS OF DISORDER

Despite what many think, anorexia is not a slimming disease. Although the image of obsessive young girls agonizing in front of the bedroom mirror and comparing themselves to the latest emaciated celebrity is a popular one, the reality of this awful condition is far more complex. Within eating disorders, food and body image are smokescreens for deeper issues. Sufferers are driven by anxiety and social pressure, not so much unhappy with their body as using it as a desperate way of expressing their unhappiness. Eating disorders are about the need for control, taken to extremes by young people unable to articulate the complex psychological problems they are facing as they advance into adulthood. Many speak of turmoil created by the existence of a hypercritical voice inside, castigating them should they stray from a predefined set of rules. Sufferer Emma says, "People with eating disorders tend to be drawn to anything rigid and rule-driven, especially when it seems pure and clean. Anorexia is about restraint, denial, and perfection and when you are already in that mind-set you are drawn to things that legitimize it."

And here lies the problem. Diets, restrictions, and food rules drive eating disorders, and the more these messages become prevalent in the media, the more fuel is poured onto the fire. It is perhaps a little simplistic to attribute this to recent dietary trends. Speaking of diet trends, Eric Johnson-Sabine, a consultant psychiatrist with many years' experience specializing in eating disorders, says:

> It is of course nothing new. It has always been like this. There is a long history of fad diets from the F-plan, Atkins, the grapefruit diet, the Beverly Hills diet. Of course many of these can lead to inappropriate solutions, but you need to be careful not to write them off. If you are judgmental with patients, they can easily become disengaged. They tend to like working with dietitians who are flexible and pay attention to the media rather than presenting an inflexible view.

Maintaining that engagement is a fine balance for those working in the treatment of eating disorders. Many I have spoken to have to hide genuine anger at the irresponsible nature of many media figures promoting rule-driven restriction under the veil of health. Messages from influencers and bandwagon-jumping celebrities, regarding both body image and dietary choices, can be extremely damaging. Renee McGregor is a dietitian specializing in the treatment of eating disorder patients. I asked her to explain what is going on, and she responded:

> Rules make people feel safe and in control. There is a feeling that if they break the rules the world will collapse. It is hard

to make people understand that it is okay not to eat raw food every day, to eat some carbs occasionally. It makes it very hard to change the mind-set. I could be making real progress with someone, then they will come back and tell me that some blogger says I shouldn't eat chickpeas because they are full of carbs. Progress would be so much quicker if this stuff didn't exist. It might not cause it, but it does keep people there. It keeps them in disordered thoughts and reassures them that not eating certain things is okay.

Most dietitians working with eating disorders share this view—that although messages of restrictive eating do not actively cause eating disorders, they do slow recovery. Any associations of purity and disgust are likely to have a particularly damaging effect on those who are already vulnerable to essentialist beliefs about food consumption. An anonymous sufferer told me:

I was in the midst of really tough therapy, trying to force myself onto three actual meals a day for the first time in a decade, starting to relinquish one by one the immense set of rules I'd amassed through the years and trying to keep to the meal plans my therapist had set. This all felt almost impossible while hearing my flatmate bang on about "guilt free" baking, measuring out minuscule portions of lentils and waxing lyrical about the latest "clean" fads. I started purposefully staying out of the flat or hiding in my room when she was cooking and then shamefully sneaking out to cook for myself, then eating in my room, sometimes in tears. If she ever caught me or asked me an innocent question about what I was cooking I had to fight the urge to throw it all in the bin straightaway . . . And I am a scientifically minded, bullshit-allergic cynical pedant who knows in her mind this stuff is hateful bilge. Such is the power of the monster, and the power of the rhetoric.

Perhaps surprisingly, the modern health movement's obsession with food-image sharing also plays into the nature of disorder. Depriving the body of food, or cycles of binging and purging, are known to drive food obsession, with many sufferers dedicating huge amounts of time to baking, cooking, food television, and magazines. The idea of spending several hours creating a superfood salad, carefully arranging and rearranging expensive, exclusive ingredients to create the perfect Instagram shot, is at the heart of the clean-eating ethos. Tragically, it also holds huge appeal to many disordered eaters.

It is probably fair to say that many well-known clean-eating stars show signs of disordered eating patterns and a troubled relationship with food, with some openly admitting that they have suffered from conditions in the past. It is also true that the democratization of media has led to many smaller influencers in closed social groups having a damaging influence. You do not have to look far to find people proposing restrictive diets as a cure for eating disorder, or groups sharing tips on conquering hunger and hiding troubled eating patterns from view.

DIETS AND DISORDER—A LINK?

Does dieting lead to eating disorders? It is hard to say, although a number of studies have shown strong links between the two.

A 1999 Australian study showed that dieting was the most important predictor in the development of eating disorders.[1] Others have shown the influence of many factors, including restrictive diets and social pressure to be thin.[2]

One thing is certain: Eating disorders are unique among psychological disorders in the extent to which they are shaped by our society and culture. Although anorexia was first categorized in the United States and Europe around 1870, for a hundred years it was considered a rare and unusual condition. It has exploded in frequency since 1970 and risen almost unabated since. Bulimia's rise is almost more perplexing, virtually unknown before the 1970s but by 1980 far more common than anorexia. Although these gains can be partly attributed to increases in understanding and diagnosis, it is thought that they are strongly linked to the development of the consumer economy. Around the world only westernized economies show significant incidence of disorder, societies where the importance of personal satisfaction has overtaken a focus on collective goals.

With the development of consumerism comes a change in women's roles within society, and in the Western world this has led to an increasing crisis of female identity.[3] Often this will drive turmoil into the lives of pubescent girls increasingly conflicted between expectations of achievement and the traditional demands for submissive dependence, all exacerbated by divergent pressures on their physical appearance.[4] Impossible expectations are being placed on the shoulders of young girls as they develop and grow.

These pressures are closely linked to issues of weight. In the United States and Europe there have been rapid changes in perceived female body shape ideals, becoming increasingly thin and less curvaceous since the beginning of the 1960s. Despite significant advances in other areas of equality, stereotypes linking curvaceous figures with low intelligence still persist. The modern ideal is one of denial and the ability to control. Similar changes have been observed in different countries around the world as economies develop and women's standing increases, with feminine status increasingly signaled by restraint and discipline rather than outward expressions of fertility.

The more an economy progresses, the more female physical ideals become infantilized and controlled, expressing a seeming fear of fertility and womanhood. The media is also inclined to sell the plasticity of the female body as a commodity, showing how it can be morphed and controlled to meet any ideal and implying that physical imperfections are representative of flaws of character. There is a great irony here: As societies become more and more individualistic there is an increasing desire to conform—with diet, exercise, and increasingly with surgery, sold as ways to meet these expectations.

Within this climate it is perhaps not surprising that as some young girls develop into adulthood, they can develop fear and disorder, thrown into turmoil by society's shifting expectations, driven by a pervasive fear of their advancing womanhood. Although there are thought to be strong genetic links that predispose individuals to disorder, the cultural factors that accompany shifting female roles in society are a huge factor. The increased prevalence of eating disorders in groups of Egyptian students studying abroad,[5] Greek immigrant populations in Germany,[6] and black South Africans in the post-apartheid era show that changing female roles and the media play a huge role in the development of eating disorders. Although genetic factors may give someone the capacity for this kind of disorder, it is our culture that sparks it off, a sad by-product of westernization and progress.

The influence of the media is also known to lead to the development of a number of weight- and body image-obsessed subcultures and cliques, which significantly contribute to body image issues and disturbed eating patterns.[7] So, although there is no causal link between dieting and disorder, there is plenty of evidence that the media obsession with weight, body image, purity, and control has a huge influence on its expression and development. To claim that disorder is genetic is to misunderstand the huge importance of our culture in the phenomenon. Why, after all, is it so common specifically in young women in this place and time, if it is all down to our genes?

Laura Dennison, who first showed signs of disordered eating at sixteen, says:

> At the time it didn't really occur to me how damaging the constant media affirmation that "skinny women are the best women" had on my self-worth until it was too late and I was in an impossible cycle of bingeing and purging; often up to five times a day. As I grew older and clean-eating and restrictive dieting became normalized, I bought every healthy cookbook going, highlighting and rehighlighting each page, unaware that this was acting as a beautiful disguise for my eating disorder. Erm, I was being healthy now! I once spent £30 [the equivalent of $40] on ingredients for one green smoothie. What a fucking waste of time, money, and energy that was.

OUR OBSESSION WITH WEIGHT

Another factor that pains many professionals working in the field is the recent rise in obesity levels, and more significantly society's increasingly troubling attitude toward it. Obesity has become recognized as the health issue of the age, and throughout the media it has become heretical to criticize any messages that might encourage weight loss, even if those messages come at a cost.

Condemnation of the obese has become accepted in our modern world, and campaigners openly launch attacks on people because of their physical characteristics, ascribing moral qualities to their outward appearance. Writer Malcolm Gladwell recently described prejudice against the overweight as "the justifiable prejudice of our age, in the way that gender was once. Those kinds of physical characteristics—what are felt to be the correlates of character traits—that's the next wave of discrimination."

Certainly much of the language surrounding obesity in the media would be deeply unacceptable if applied to other physical characteristics such as skin color or disability. We are forever told that the fat are to blame for a devastating health crisis, that their health care should be withdrawn, that they are weak, ugly, a blight on society, costing us billions because of their slovenly ways. Weight has become a moral issue, riven with deeply embedded class prejudice. As weight increasingly correlates with lower socio-economic groups, a rich and skinny elite moralize at the obese, feckless poor, offering them pearls of wisdom to cure their dietary ills.

The more that the overweight are shamed, and the greater the gap between average and ideal becomes, the more body-weight anxiety is created.[8] That anxiety breeds disorder, focusing upon the achievement of an ideal physical form and an association between weight loss and moral virtue. Although for the most part the intention is to help the overweight, it can be extremely damaging for people with eating disorders. The laser focus on obesity provides a constant battle for those involved in the treatment of these conditions. Jane Smith, who runs the charity Anorexia Bulimia Care (ABC), says:

> I think that obesity can overshadow the problem of eating disorders and their treatment options. For us at ABC it hinders receiving funding because anorexia nervosa and bulimia nervosa are often viewed as issues of teenage vanity and not as important a risk to health as being overweight. The lowering of the BMI [body mass index] Scales for one thing gives people who may become underweight or anorexic no safety net. BMIs of eighteen were always

considered underweight but now they are part of the normal range.

Our obsession with dietary health has created a world where the end point of all our eating is seen as the creation of a perfect physical form. The idealized photographic image is the center of our world, aesthetic appearance confused with notions of health, the attraction of an Instagram shot a signaling of our morality. Food should be so much more than a medicine of weight control. Culinary joy should be free from rules, restrictions, and guilt. No one should ever be made to feel guilty about food, no one should feel shame over cake, a sandwich, a bar of chocolate, or a bag of candy. All food can be embraced because all food is healthy in some way and everything can form part of a balanced diet. It is only our obsession with health that has made things seem so different. In my many discussions on this subject, and despite all the deeply distressing, heart-wrenching stories of loss that have been relayed to me, the comment that has stuck with me the most was an offhand remark from the dietitian Renee McGregor, a frequent Angry Chef collaborator. She said that if you are going to a barbecue at someone's house tomorrow and are worried about what you are going to eat, then that is probably a sign of disordered eating.

I am sure that most of us know people who fit into this category, scared at the prospect of eating a cheap burger or some processed white bread. But if we analyze this fear it is beyond logic, because unless we believe that eating a one-off meal at the house of a well-meaning friend will make us sick, or fat, or permanently damage us in some way, then we have nothing to fear. These thoughts of toxicity, sickness, and danger are driven by the media, diet books, clean-eating blogs, detox gurus; by sugar-shaming, fat-phobic bigots, selling us lies and ascribing morality to perfectly normal dietary choices. No food should be feared, no choices deemed "wrong." We should be free to embrace the huge variety that the world of food has to offer us, not restricted in our choice based on the moral values and pretentions of others. The end goal of all eating should not be a good-looking Instagram shot. The pleasure of eating should be embraced for what it is: variety, joy, precious moments shared.

It is worth bearing in mind when you see anyone restricting their food choices and telling you to do the same that an estimated 20 to 25 percent of people with disordered eating will go on to develop a full-blown eating disorder. Although there is unlikely ever to be a causal link established between clean eating, fad diets, wellness regimens, and the healthism that pervades modern dietary choices, that does not absolve them from blame. Those in the media have a responsibility to spread messages that will not do harm. Restriction and rules around food can harm a huge number of vulnerable people. Just because someone's problems are driven by deeper issues, this does not mean that they are fair game to be exploited by messages that prey on their fears.

Shaming anyone for their physical appearance is never acceptable, no matter what your intention. No body shape is ideal; no one should be led to believe that there is a physical perfection they can achieve. We have a right to feel concern for those whose weight issues carry a danger of physical harm, but we must never judge them for their appearance or make implications as to their character because of it.

I have seen the damage done by clean eating, by media-driven food obsession and by anxiety around body image. I have met people who recovered and I have spoken to families of those who did not. I cannot accept that the problems of obesity are so great that we should continue down the path of weight prejudice and healthism. I will fight to my last breath against irresponsible and harmful food messages in the media, because they contribute to a hidden plague that shames us all for allowing it to grow unchecked. These messages might not cause disorders, but I have no doubt that they cause great harm.

FOOD AS JOY

I have devoted my life to helping people enjoy food in the hope that it will bring them as much pleasure as it does me. Compared to the many astounding, brave, and dedicated people I have met in the course of writing this book—the caretakers, researchers, professors, doctors, dietitians, and campaigners—my strange life choices seem trivial in comparison. But when it comes to eating disorders, to people struggling so deeply in their relationship with food that they have lost what it means to feel joy, then my life's mission makes sense. I will fight to engage people with food, to eat a varied diet, to seek pleasure without needless restriction. I will show the nature of food as celebration, as a glue to cement our most important bonds, as something to be loved for what it is, the creator of memories and bringer of joy. The moment that messages of restriction appear, I will tear them down the only way I know how: by exposing the liars and charlatans and shining a bright light upon their quackery. But more than that, I will spread messages of culinary love, encourage everyone I know to embrace the rich variety that the world of food has to offer, and strive to cook food that makes the heart sing.

We have created a society where clean eaters, health obsessives, body shamers, fattists, and anyone peddling restrictive diets are seen as virtuous and good. It is time they were revealed for what they are: shallow, prejudiced, irresponsible, and wrong. These things need to change, because if they do not, innocent, vulnerable victims will continue to fall. Our society is driving eating disorders, and so societal change can drive them away.

If we ever reach a point where food has no rules, no needless restrictions, no obsession with body image or constant moralizing, then I imagine that there will still be obesity and eating disorders. But I am also fairly sure that in a world where people enjoy food for what it is, and embrace all that it has to offer, it will be far harder for obesity and eating disorders to take hold.

Part IV

THE DARK HEART OF

PSEUDOSCIENCE

Chapter Sixteen

RELATIVE RISK

N ow, I need to go back to the instinctive brain, because we are going to talk about some of his favorite things: fear, outrage, and risk.

Risk. That sounds like statistics. You know I don't like statistics. You haven't been talking to Norm Spiegelwhatsit again have you?

Maybe. And it's Professor David Spiegelhalter, who cowrote *The Norm Chronicles*.

Yeah, him. I like him and everything. It's just statistics. Bleugh.

There will only be a little bit of statistics. I promise I won't make you do any calculations.

FEAR, OUTRAGE, AND RISK

We are not great at understanding risk. The main problem is we are easily manipulated. News headlines, images, social media, and the company we keep all have an influence on how we perceive danger. The groundbreaking work of psychologists Amos

Tversky and Daniel Kahneman in the 1960s and 1970s showed that people do not judge risk based on a careful assessment, but act only on the limited amount of information available to them. This is known as the availability heuristic. It means that if the newspapers are full of powerful and emotive stories about terrorist attacks, these will be the images most accessible to our minds when thinking about risk, and so we will be likely to overestimate the danger of being affected by such an incident. As a result, we may decide to travel by car rather than airplane, despite the far greater chance of encountering injury or death on the road, purely because the terror of an aircraft hostage situation comes more readily to our minds.

We also have a tendency to confuse fear with outrage. Although many people die in car accidents, for the most part these are just random occurrences—awful and terrifying, of course, but generally hard to blame. The same is true of many accidental deaths and chronic diseases—they fall under the "bad shit happens" category, something that sits uncomfortably with our love of cause and effect. However, when deaths are associated with something that we find morally unacceptable, such as terrorism, child abduction, chemical leaks, environmental destruction, or corporate greed, our outrage is expressed as a greater sense of fear, and these examples become more prevalent in our minds when we consider risk. Although for the most part these morally unacceptable risks are less statistically significant, often people live their lives paying far more attention to them than to greater risks found elsewhere.

The world of food has a number of similar problems. Newspapers are full of stories about the risks of consuming various foods, often written with sensationalist anecdotes. For something to be in the media, it needs to be unusual, so our perception of risk is skewed by controversial opinions and ideas. The old, boring, non-newsworthy scientific consensus is rarely reported, leaving the media limelight for self-styled mavericks with a book to sell. We have seen examples of this a number of times so far, with sugar being revealed as a toxin, saturated fat becoming a superfood, our body being acidified by chicken, and gluten being declared unhealthy for all. These are niche beliefs not held by most of the scientific community, but because they are the stories in the news, they are more available in most people's minds and often drive our choices about food and health.

Outrage, too, plays a part, as we have seen extensively in the examples covered. Nothing makes better headlines than a conspiracy or a cover-up, with the mysterious forces of big business colluding with experts to harm the public. The outrage we feel at the idea of a contaminated food supply, or corrupt scientists producing false dietary advice to fuel corporate greed, is expressed as fear and disproportionately alters our decision making.

The scientific community is left with a difficult quandary, because engaging with those holding false beliefs brings them into the debate and draws attention to them. But to ignore them when they appear in the media so frequently means that these falsehoods will often be the only bits of information readily available when people are making decisions.

Similarly, the popularity of bestselling books outlining a new approach to diet, health, or weight loss will make that particular concept of risk or benefit jump to the front of our minds, giving it an unjustified importance. The bestseller lists are full of dramatic and sensationalist titles outlining why gluten is evil, or why sugar is filling your brain with toxic fog, but rarely will you find anything repeating the consensus views of the scientific community, because those views are not as snappy: "Prevent Long-Term Weight Gain by Not Eating Too Much"; "Avoid the Damage Caused by Excess Sugar Consumption by Eating Less Sugar (Providing You Currently Eat Too Much Sugar—If You Don't, Just Carry On as You

Are)"; "The Modest Long-Term Benefits Associated with Limiting Your Intake of Saturated Fatty Acids."

Maybe a chef could write a book like that? He could put in some barbecue tips to make it more interesting.

Given all the potential for confusion, it is crazy to expect people to make decisions based on a careful weighing-up of all the statistical risks. As we have discussed already, we believe the sort of information we are generally inclined toward, especially when it is delivered by the sort of people we like to trust. These are simple shortcuts that we use to help us navigate the world, and when it comes to nutrition and health, these are completely understandable given the complexity of information we are confronted with. Although we should always stop to think and look for evidence whenever we can, for almost all of us it is nearly impossible to assess the quality of every relevant scientific study. Scientific research, especially the sort of complex epidemiological study that underlies a lot of nutrition and health advice, is impenetrable to all but a few people. We all have to rely on experts to interpret it and communicate any important findings to the world. This comes with a number of problems, as David Spiegelhalter points out:

> Although it is impossible for people to make decisions based on weighing up probabilities, they are actually pretty good at understanding risk. The problem is they are often let down by the people communicating it. They are exposed to badly packaged statements by people wanting to manipulate them. There are certain ways of framing information used by people who want to scare or reassure the public.

Clearly in the world of nutritional pseudoscience, there are people with vested interests in trying to scare or reassure us—and information is often framed in a certain way to fit an agenda. But within the world of science, there are sometimes subtler reasons at play for this sort of miscommunication.

I thought this book was supposed to be about pseudoscience. You've been going after the real scientists a bit, haven't you?

It is true that I detest nutritional pseudoscience most of all, but this book is about general misinformation in the world of food. If there was more effective science communication, it would make it far harder for pseudoscience to flourish, so understanding these issues is very important. As we have already discussed, there are many reasons why science gets communicated badly, with a number of parties (researchers, university publicity departments, journalists, charities, campaigners) having a vested interest in building up claims and seeking publicity. But surely when science deals in hard facts, there is only so much misinterpretation that can occur?

I feel some statistics coming on.

RELATIVE AND ABSOLUTE RISK

Key to understanding why some nutrition research is communicated badly are the concepts of relative and absolute risk. When complex epidemiological studies are conducted into diet and health, the outcome tends to be one of *relative* risk. Studies will look at the differences in risk levels between two groups and investigate how certain behaviors will affect the outcomes.

Huh?

Let's try an example. A well-publicized study from a few years ago showed that consuming bacon was a risk factor in developing colon cancer.[1]

Noooo! Not bacon. I love bacon.

Yep, afraid so. I am simplifying the figures here for illustrative purposes, but the study showed that the consumption of cured, processed meats significantly increased the risk of developing this horrible disease. Studies of populations showed that groups of people who consumed an average of fifty grams (just two pieces!) of bacon per day over the course of their lives had an 18 percent greater chance of developing colon cancer than did groups that did not.

Eighteen percent!!! I'm scared. I will never eat bacon again.

Now, I know I have been trying to drum into you about correlation, causation, and the potential for confounding factors, and I am sure you can imagine that the sort of people who consume 50 grams of bacon every single day might not be the healthiest specimens, but the study used some clever techniques to make sure that the correlation is indeed real. Colon cancer is a very serious, extremely unpleasant condition, and there is actually very good evidence that bacon, ham, and other cured, processed meat consumption has a strong link to its occurrence. The case for bacon does not look good. Although delicious, is it really worth increasing your risk of cancer by 18 percent?

The 18 percent increase is a measure of relative risk, and as I mentioned this is the output of most epidemiological studies. High processed-meat consumption increases the risk of cancer by 18 percent in relative terms when compared against groups who eat less processed meat (with any potential confounding factors carefully adjusted for). But what does this number actually mean?

For everyone reading this book, whether you eat bacon or not, the risk of developing colon cancer at some point during your life is fairly high, around 6 percent. If you eat fifty grams of bacon every day (which is a lot of bacon—a family of five consuming this amount would be eating 3.85 pounds [1.75 kg] per week), that risk will increase to about 7 percent. This figure, the increase from 6 to 7 percent, is known as the *absolute* risk. This is an increase of only 1 percent, yet when it is compared against the original risk, it becomes 18 percent greater.* There is definitely a link between colon cancer and bacon, and the relative risk does increase by 18 percent, but in reality the study shows that people who eat a lot of bacon every day will increase their absolute risk by 1 percent, which, although serious, does not seem quite as bad. It is easy to see how the relative risk is likely to worry people more than the absolute risk.

There are also other ways of framing the same information that might have an effect on our decision making. If I say that eating bacon increases your risk of colon cancer by 1 percent, that still sounds like a significant increase for just one ingredient. But if I say that the chance of avoiding colon cancer is 94 percent for people who do not eat bacon and 93 percent for people who enjoy huge amounts of bacon every day, the odds definitely sound better. And if I say that out of a group of a hundred people, if they all eat a massive bacon sandwich every day for the rest of their lives, then one more of them will develop colon cancer, then perhaps the risk seems less significant still.

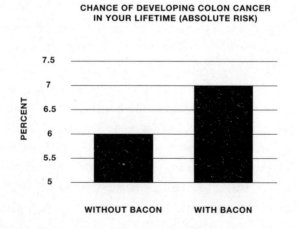

CHANCE OF DEVELOPING COLON CANCER IN YOUR LIFETIME (ABSOLUTE RISK)

* I know that 1 is not exactly 18 percent of 6, but when the exact figures are used without rounding off, the increase is 18 percent.

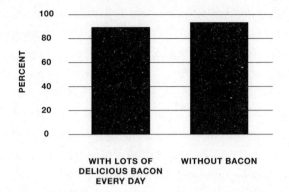

**CHANCE OF AVOIDING COLON CANCER
IN YOUR LIFETIME (ABSOLUTE RISK)**

Yay. Bacon is good.

No, bacon is not good. There is strong evidence that eating a lot of it does increase your risk of colon cancer, a serious disease that kills lots of people. But it is clear that framing it in different ways can change the way that we think about risk.

So, which one is right? I'm confused now. I preferred it earlier when I was just scared.

All are correct, but lots of psychological studies have shown that absolute risk is the best way of communicating if you want the public to understand you.

Ah, good. So, everyone uses absolute risk then.

Unfortunately not. Although it is well known that absolute risk is the best way of communicating, it is rarely the one that is used. This is perhaps because it does not generally create the biggest headlines. The colon cancer and bacon link was communicated around the world as a dramatic increase in relative risk, and although the research was significant, many might consider it unfair to frame it in a way that is known to confuse the public and create unnecessary fear.

Why does this happen?

Perhaps the main reason for this is that relative risk is the final output of most epidemiological research. Researchers rarely provide information on absolute risk, as producing this is not the intention of their work. It is not the job of the researchers to translate what they do into information for public consumption, and so this is something that is rarely done. It could be said that this sort of translation is the job of science journalists, but generally this is beyond their capabilities. Perhaps it should be the job of university publicity departments. A recent *British Medical Journal* paper did lay the blame for misleading scientific reporting firmly at their door, but in reality a lot of the complex work of translating research findings into absolute risk figures that the public will understand will be beyond them, too. As David Spiegelhalter explains:

> The public is often ill served and given an overexaggerated view of something's importance. Science is not written for the public, and the process of turning it into good information is not the job of scientists. Scientists, press departments, and journalists all want a story and all have a tendency to indulge in exaggerated or misleading claims. Everyone has an incentive not to look at the bigger picture.

Thankfully there are places we can look to for good advice and a number of people (like Professor Spiegelhalter) who are attempting to address these problems. The National Library of Medicine (part of the NIH), the Department of Health and Human Services, and the CDC all work hard to try to dispel myths and offer information that helps the public understand risk and make informed choices. But the sensible voices are not the loudest.

When science is communicated badly, in ways that are known to mislead the public and result in their making poor decisions, it is hardly surprising that problems and confusions arise. There needs to be a greater consensus about how risk is communicated and more responsibility taken by all parties when presenting the results of scientific research. Until there is, you should learn to spot relative risk when you see it, and be wary of the power it has to influence your decisions.

THE RISKS OF BACON

Although a number of campaigners have used the link between cured meats and colon cancer as evidence of the dangers of nitrates added to processed meats, there has been no causal link established between the two. Given that nitrates are added to prevent the occurrence of deadly microorganisms—and without them, bacon would not actually be bacon (they make it pink)—I personally do not think their vilification is justified. Processed meats contain a lot more stuff than just nitrates. Blaming one chemical is an unscientific jump that fits the convenient "chemicals are bad" narrative. There are a number of theories, and one day I might be proven wrong, but until there is good evidence, this sort of leap is unacceptable.

Also, sometimes we do need to accept risk. Nitrates are added to bacon to inhibit the growth of *Clostridium botulinum*, a bacterium that produces a deadly nerve toxin. It is not always possible to remove risk entirely from our lives, and sometimes we do need to judge which risks we want to take. In many cases this can be simply calculated, bacon versus botulism, for instance, or flying versus driving, but one of the most important things to consider is that risk sometimes needs to be weighed against pleasure. Although some might dispute this, bacon sandwiches are not essential to life. They are, however, capable of bringing about fleeting moments of great joy, and so that joy must be weighed against the potential long-term risks. Should we eat bacon every day? Probably not. If we want to eat a varied diet, we should probably avoid eating anything every single day. Can we afford to enjoy the odd bacon sandwich on a Sunday morning? With a little bit of butter and brown sauce? I know what I think.

RISKY PROSPECTS

Even when we are fully aware of the risks, sometimes there are still problems. These problems are especially prevalent as we come to the part of the book where we look at the worst bits of pseudoscience.

Oh. That sounds serious.

Some of it is. That is why, before we start, we need to talk a little more about risk and decision making under stress.

More statistics?

No, not this time. We are going to play a game.

Hurray. I like games. What game are we going to play? I hope it's Hungry Hippos. Or KerPlunk. KerPlunk is fun.

Well, I am going to let you decide which game we play. You need to tell me which one sounds the best.

KerPlunk. I choose KerPlunk.

No, I am going to give you two options and you must choose one of them. Ready?

Yep.

OK. The first choice is a game where you have to give me $500. And that's it.

What? Bullshit. What a stupid game that is. Not playing. What's the other choice?

The second is a game where you throw dice. If you throw a six, then you don't have to give me anything. If you throw any other number, you have to give me $700.

Oh. Is that the only choice? What happened to KerPlunk?

KerPlunk was never on offer. Which do you choose, the first game or the second game?

The one with the dice, I guess. At least I have a chance then.

OK, good. Want another try?

What do you mean?

Do you want another two options to choose between?

OK. I hope they are better this time.

I think you will like them a bit more. In the first option this time, I give you $500. Once again, that is it.

Yay. That sounds fun. I like being given $500. What's the second choice this time?

In the second game you throw dice. If you throw a six, you get nothing. If you throw anything else, you win $700.

Oh. That's good, too. But I prefer the first one. That way I definitely get $500. I could buy a gold KerPlunk set. And some luxury chocolates. The ones individually wrapped in foil.

Good. Thank you, instinctive brain. You have done a good job of explaining a little bit about people's attitude to risk.

In 1979, Kahneman and Tversky published a paper entitled "Prospect Theory: An Analysis of Decision Under Risk." It became one of the most influential and important works of its time, in many ways creating the field of behavioral economics. Broadly speaking, prospect theory describes the way people make decisions that involve risk, specifically looking at how these decisions are often seemingly illogical.

People behave in different ways when outcomes are framed positively or negatively. In games similar to the one we have just played with the instinctive brain, it seems that when people are faced with negative outcomes, they are far more likely to take the risk of losing more. When all the outcomes are positive, however, especially when one of the games is a sure winner, we are less likely to take a risk, even though it concerns the same amount of money. In fact, in experiments not unlike our simple one, 92 percent of people choose to "throw the dice" when faced with losses, yet only 20 percent of people do so when faced with gains.

When we look at the choices, this seems obvious. When we are definitely going to lose $500, it seems worth taking the risk of losing an extra $200. You are already a loser, so being a slightly bigger loser does not really make much difference. In for a penny, in for a pound. But when we have already won $500, the prospect of winning the extra $200 is not very enticing and not worth the risk. In this case, a bird in the hand is worth more than a slightly bigger bird in the bush.

In the context of gains, the extra $200 does not seem worth the risk, but when it comes to losses, the risk of losing exactly the same amount is somehow worth taking. It will still have the same impact on our pockets, but we will not feel it as much emotionally. Prospect theory is full of these conundrums, things that make perfect sense to our instinctive brains, but no sense at all to economists. However, it is this finding in particular, the fact that we are more open to riskier outcomes when all our options are negative, that has particular relevance in this section of the book. This is especially the case when we come to chapters 17 and 18, areas where people facing difficult situations and tough choices are susceptible to false hope.

Much of what we have discussed so far concerns people's susceptibility to false belief relating to vague new age terms, such as "achieving holistic wellness" or "getting the glow." In some cases, ill-defined and hard-to-diagnose syndromes, such as chronic fatigue or multiple chemical sensitivity, are exploited because of a lack of conventional treatment options and the unreliable self-assessment of symptoms. But true quackery has an even darker side. The worst examples tap into this aspect of prospect theory, exploiting the knowledge that when people are in tough situations, when all the possible paths look treacherous, they are more inclined to take risks.

Life does have a habit of dealing us some pretty awful cards at times. It violates common sense when bad things happen to good people. Serious, terrible conditions, such as cancer, AIDS, motor neuron disease, or Alzheimer's affect millions every year, and there is nothing like a devastating, life-changing diagnosis to change the way we perceive risk. When the chips are down, we are inclined to take bigger and bigger chances, pinning our hopes on a course of action we would not dream of taking in more positive times.

It requires a special kind of evil to offer false hope to the sick and vulnerable. It is beyond me how someone could exploit our inclination to risk at the darkest point in our lives for financial gain. But as you look into this darkest of worlds, it becomes clear that the reality is rarely that clear cut. Although there are doubt-less some evil charlatans, there are many more who misguidedly believe that they are helping others. People who think that the options they present are somehow kinder, gentler, wiser, and more benign than those that they replace. This is perhaps the most tragic thing of all. Many who offer false hope are just as deluded as those who follow them, and some believe the nonsense that they preach to such an extent that they will take the greatest risks imaginable.

In the next few chapters we explore a few of the very darkest places and remind ourselves that, although it is easy to dismiss much of the nutritional nonsense we have looked at as "harm-less" or "just trying to get people to eat healthily," once you start scratching at the surface, it can reveal the blackest of hearts.

Chapter Seventeen

THE GAPS DIET

I was often found asleep on the floor next to my son's bed for the first fourteen years of his life, only to have to get up, get my other child ready, and go to work myself . . . Imagine the constant screaming that comes from your child being unable to express their needs. And when language comes, imagine your child sobbing, "Mum, help me, I can't switch my brain off," or repeating the same segment they have watched on a DVD over and over again in an attempt to self-soothe. Imagine the looks and stares when your child is having an Oscar-winning tantrum because they have changed the packaging of his favourite yogurts, or laughing hysterically when a baby falls off a ledge because they made a funny face.

—JENNY (PARENT OF AN AUTISTIC CHILD)

W e are all prone to make mistakes when under intense stress or pressure. As we discussed earlier, in situations when all potential outcomes are bad, we often become inclined to riskier options. There are also many studies showing that in stressful situations we are likely to change the way we make decisions, often allowing our instinctive brains to take control.[1] French psychologist Jean Piaget described how in difficult times adults can regress in their thinking, moving back to a more childlike state and a belief in magical ideas. When times are tough, our protective adult cynicism can be stripped away, leaving us willing to believe almost anything.

Certainly the parents of many autistic children face a day-to-day existence full of constant and intense stress. It has been described to me as being similar to the first fretful few weeks of parenthood, with sleepless nights, regular screaming fits, frustration, and overwhelming exhaustion, but magnified, and lasting somewhere between ten and fourteen years. As an autistic childhood progresses, many parents are left coping with a physically developed child with little sense of danger and a worryingly abnormal view of social interactions. Although every parent will probably have to deal with a few tantrums, the parents of a severely autistic child might regularly have to cope with the prospect of their punching another child to get themselves out of a situation, or pulling down their pants and pissing on the doctor during a regular checkup. This can create the sort of prolonged stress that few of us will ever have to deal with, compounded by a distinct lack of public sympathy, hard as it is for most of us to distinguish autistic behavior from delinquency. Autism is a mysterious and poorly understood condition, covering an extremely wide spectrum, from individuals with Asperger's syndrome requiring little specialist care right through to the severely disabled, without language, and unable to care for themselves into adulthood. It is incredibly hard to define and confounds most of us, leading to admonishment and guilt being heaped upon already desperate parents.

It is perhaps not especially helpful that, although things are improving, for many years autism has been a diagnose-and-dump condition for many health-care providers. Having battled to get a diagnosis, parents are often simply sent away with a "good luck." To make matters worse, scientific understanding of the condition is still riven with uncertainty. Although it is known to have an extremely strong genetic basis, it increasingly seems that there are a number of different factors acting in tandem.[2] Despite many years of study, no single mechanism explaining all aspects of the disorder has been found, and true autism research progresses only by accepting this complexity. With no potential unified mechanism or cause, it is unlikely there will ever be a single defined cure or universally effective intervention. Appreciating this reality is seen by many to be the key to making progress in treatment and care.

We should also consider that autism is not necessarily a condition that requires a cure, but may simply be a different way of functioning. A number of groups have formed in support of autistic rights, suggesting that perhaps understanding and acceptance is the most powerful way of dealing with the condition. This is an interesting viewpoint and not without merit. For numerous reasons I have contact with a number of autistic individuals through the world of Angry Chef, helping with research and giving a valuable perspective on the issues I regularly write about. In my limited contact, although I am aware that their condition affects several aspects of their lives, the prospect that they somehow need curing seems bizarre. That is probably a discussion for another book, but it does underline some of the complexities of the condition—the fact that it is not one clearly defined set of symptoms or causes, that treatment is likely to require a number of approaches, and that increasing public acceptance and understanding are perhaps the most important strides we can make.

What is important for this chapter is that for the parents of many autistic children, things can be very fucking hard. The constant long-term stress many parents are under has the potential to thoroughly scramble their brains. Jenny, the parent quoted at the beginning of this chapter, also told me that the two groups of people that it is easiest to sell something to are the parents of autistic children and balding men. Parents are desperate and looking for any way out of the grinding pressure. They are susceptible to anything that might promise some respite, anything that may give them some hope, any port in the storm.

When it comes to autism, conventional medicine offers little hope. At best it suggests long-term treatments that might provide slow, incremental improvements over many years. Given the huge pressures many parents are under, combined with the scientific uncertainty about causes and treatments, it is hardly surprising that autism has become a hotbed of often dangerous pseudoscience. From the persistent and damaging link to vaccines through to abusive "cognitive therapies" and damaging chemical "treatments," there are a number of people who appeal to the desperation of well-meaning parents with promised universal cures. There are also many myths surrounding diet, and even more concerning the effect of unspecified toxins in the environment.

Unproven, ineffective, and sometimes dangerous treatments are forced onto desperately vulnerable children with the potential to do lasting physical and psychological damage. Often these treatments are administered by primary caregivers with the best of intentions, prompted at times of great vulnerability into harming those whom they love.

Richard Mills, the research director of the UK charity Research Autism, is an outspoken critic of the many unproven treatments being offered:

> If you speak to autistic adults, many will tell you that large parts of their childhood were very tough due to often abusive treatments being forced on them. The treatment of autism should not be about trying to cure, but trying to help, and whilst parents are going down these roads, children could be in programmes that work with their autism,

leading to a genuinely happier child. Unfortunately, there is still the belief that anything can be done to an autistic child, even hazardous and cruel treatments, with no risk of prosecution. There really needs to be better legislation, but the state still takes a view that the lives of severely disabled children are somehow less important, and that anything is acceptable if it promises a cure. The reality is that no treatment has been shown to cure or reverse autism. Some of the people offering treatments may be well meaning, but others are quite cynical. The truly wicked thing about it is that it falsely raises hopes and expectations.

NATASHA CAMPBELL-MCBRIDE

I first came across the GAPS diet by accident while writing a piece about the Hemsley sisters, our favorite purveyors of clean eating and coconut oil. I was picking through some familiar themes when I noticed that some of their recipes were listed as being suitable for the GAPS diet, a term that I was previously unaware of. A little research revealed a troubling and mysterious movement.

According to the founder, Natasha Campbell-McBride, GAPS, which stands for "gut and psychology syndrome," is a condition that underlies many different medical problems and can be treated with a specially designed dietary intervention. The diet focuses on the removal of all grains, refined carbohydrates, and starchy vegetables from the diet, and a concentration on meats, offal, and large quantities of "bone broth." It claims to "naturally cleanse" the liver and colon, and in doing so help treat many serious neurological disorders, including autism. The world of GAPS and its founder is a perplexing one, a rabbit hole that runs deeper than I can fully describe in this book. But as I shall show, nothing exposes more the dangers of abandoning the tenets of reason, and never is the need for evidence so clearly revealed.

Natasha Campbell-McBride is a Russian-qualified doctor believed to have received her medical degree in 1984 from Bashkir State Medical University in Ufa, Republic of Bashkortostan, when it was part of the old Soviet Union. She worked as a neurologist and neurosurgeon before moving to the UK where she studied human nutrition in Sheffield. Although not currently registered to practice medicine in the UK, she still uses "MD" in her profile and is a senior partner at the Cambridge Nutrition Clinic, specializing in the use of nutritional approaches to treat a number of medical conditions. The website of the Weston A. Price Foundation, where she is an honorary board member, describes her as "one of the world's leading experts in treating children and adults with learning disabilities and other mental disorders, as well as children and adults with digestive and immune disorders."

Given that Natasha never published any research in serious medical journals, the claim of being a leading expert is perhaps a little far-fetched, but it is clear that in many circles she is seen as an important figure. To be fair to her, although she has not practiced medicine for many years, she does have some genuine credentials, including experience as a neurosurgeon, which is more than can be said for many nutritional gurus.

Soon after arriving in the UK, Natasha claims that her son was diagnosed with autism. Receiving little hope from conventional medicine, Natasha embarked on a journey of research and discovery (although there is no record of her publishing any findings, even a case study of her son's treatment). She claims that the treatments she developed cured her son of the condition, so inspiring her to spread the word about her miraculous ideas. She set up her Cambridge clinic and quickly realized that the dietary interventions she had developed had the potential to treat many other conditions. In 2004 she published a book, *Gut and Psychology Syndrome*, in which she outlined her findings.

Natasha's beliefs concerning the treatment of autism are indeed revolutionary. *Gut and Psychology Syndrome* describes a condition that she believes explains a number of diverse mental and physical problems; a unified cause for all forms of autism, but also for a huge number of other conditions. According to her book and the FAQ section of her website, gapsdiet.com, GAPS is behind all autistic spectrum disorders, dyspraxia, attention-deficit disorder, dyslexia, depression, schizophrenia, epilepsy, various behavioral and learning problems, allergies, failure to thrive (a slightly old-fashioned description she uses for a condition in young children known as faltering growth), Tourette's, bipolar disorder, obsessive-compulsive disorder, teeth-grinding, eating disorders, gout, bedwetting, night terrors, body odor, drug addiction, and eczema. In more recent times she has also made claims on her website for the diet's ability to treat a number of autoimmune conditions, including celiac disease, Crohn's disease, and type 1 diabetes.

These are bold claims, especially given the view of most serious researchers that even autism does not have a single defined cause. It does perhaps seem a little unbelievable that medical science might have missed this one unifying reason for so many conditions, but scientific progress is often inspired by the ideas of outliers, so let's give her a fair hearing. Natasha is certainly convinced, as are many followers, and she opens her book with a quote from Hippocrates, that "All disease begins in the gut," perhaps taking it more literally than even he originally intended. To find out how she justifies her incredible claims, we need to look at what the GAPS diet involves, and what Natasha believes is actually the mechanism for so many diseases.

SO, WHAT IS THE GAPS DIET?

Natasha Campbell-McBride does indeed seem to believe that all disease begins in the gut. In her book, she explains that the conditions she describes are a product of toxins leeching from our gut into the bloodstream and causing sickness. She explains:

> The purpose of the treatment is to detoxify the person, to lift the toxic fog off the brain to allow it to develop and function properly. In order to achieve that we need to clean up and heal the digestive tract, so it stops being the major source of toxicity in the body and becomes the source of nourishment, as it is supposed to be. As more than ninety percent of everything toxic floating in our blood (and getting into the brain) comes from the gut, healing it will drop the level of toxicity in the body dramatically.

Her ideas stem from a belief that toxins in our food (and generally in the environment) get into our bloodstream if we have a damaged gut, and these toxins cause a "toxic fog" in our brain leading to numerous developmental conditions. Some have wondered how she accounts for the glaring issue that the blood-brain barrier would prevent these toxins from reaching the brain, postulating that as a neuroscientist she would doubtless be aware of this, but she does not seem to cover this anywhere in her book or on her website. I must remember to ask her if I ever get the chance.

She claims that once we can heal our damaged guts—easy to do with Natasha's special diet—we will be free from toxicity and also free from autism or any of the other diseases and disorders she believes are caused by GAPS. A key point about the diet is underlined in this passage from her website: "This programme has evolved through the personal experience of Dr. Campbell-McBride's family and clinical experience with thousands of GAPS children and adults around the world." The GAPS diet has not been tested for effectiveness by any sort of clinical trial, but developed by Natasha based on her personal beliefs about what is going on.

This is a long way from the evidence-based medical interventions she should have been trained in at medical school. Exactly how she developed these beliefs is still unclear, but a clue is revealed when we look at the diet's foundation.

The GAPS diet has its origins in the specific carbohydrate diet (SCD), created in the early 1900s by Dr. Sidney Valentine Haas. SCD is a rule-driven and restrictive low-carbohydrate diet that showed some promise in the treatment of celiac disease in the days when that condition was poorly understood. In 1958, a few years before Haas' death, he treated the daughter of a woman called Elaine Gottschall, who was so impressed that she was inspired to spend the rest of her life studying the diet, writing a number of books, and evangelically spreading the word about its remarkable curative properties.

Elaine believed the diet could be used in the treatment of numerous conditions, including autism and cystic fibrosis, which presumably explains how Natasha came by it. Despite its existence for over a hundred years, no reputable studies have shown the SCD as having any effect on the conditions it is purported to cure, but still the myths about it persist—largely due to a number of powerful anecdotes and some extremely passionate advocates. I receive more abuse about my occasional criticism of the SCD than anything else (even Paleo, a diet that is in many ways similar), from people convinced that it has cured them of debilitating ailments, particularly when it comes to inflammatory bowel disease. All I can ever do in response is repeat my usual mantra that anecdote is not evidence, correlation is not causation, no studies have shown any effect, and there is a known risk of nutrient deficiency associated with it.[3] As if the SCD was not restrictive enough, Natasha added a number of additional exclusions and detox protocols, as well as a quite staggering level of complexity, creating a multistage program of dietary restriction. There are three main phases to the GAPS diet: the introduction diet, the full GAPS diet, and a coming-off stage. The one most worth focusing on is the introduction phase, formed of six stages and designed to "heal the gut."

There is no defined timeline for the introductory phase, which apparently can take up to a year. This includes an initial exclusion stage, where you are allowed to eat homemade meat stock (Natasha is absolutely convinced of the curative properties of meat stock, or "bone broth" as she calls it, and its ability to heal), vegetable soups, probiotics, and some supposedly "probiotic" fermented foods, such as sauerkraut, yogurt, and kefir. Slowly throughout the six stages, other foods are introduced, but these are few, and the diet is highly restrictive. Zoe Connor, a dietitian who specializes in autism, analyzed the nutritional adequacy of this stage of the diet, finding it to be below the recommended intake of vitamins, minerals, protein, and calories and, for anyone following it, placing them at risk of malnutrition.[4]

It is worth remembering at this point that although the GAPS diet is very much marketed as a cure-all for a myriad of diseases, it explicitly appeals to the parents of autistic children. Natasha believes that the earlier the treatment is started, the greater the power to heal, stating that "the younger the child when treatment begins, the better results I see. If you catch the autistic child under the age of three, there is about a sixty to seventy percent chance of full recovery."

Let's just think about that for a moment. This is a malnourishing diet being marketed to the parents of very young, desperately vulnerable children, perhaps for up to a year of their life. In case there is any doubt just how dangerous this could be, Zoe Connor is on record as saying, "If a child followed GAPS to the letter they could be seriously harmed or even die."

Compounding the dangers of dietary restriction, in response to questions about what to do if anyone on the GAPS diet should become dangerously underweight in the introductory phase, the frequently asked question section of the GAPS website says, "Some patients start on the Introduction Diet very thin and malnourished because their body's [sic] are unable to absorb proper nutrition." It continues, quoting Natasha:

> Regular consumption of grains and processed carbohydrates causes water retention in the body. As you stop consuming these foods, you will loose [sic] that excess water and hence loose [sic] some weight, which usually happens in the first few weeks. Without the water retention you will get to your real weight and size, which will show you the real extend [sic] of your malnutrition.

Apparently the weight loss caused by this highly calorie- and nutrient-restricted diet is actually just revealing your "real" weight. Again, it is worth reinforcing that many of the people forced onto this diet by their well-meaning parents will be young children with severe learning and language difficulties, many of them already unable to communicate their distress. Also worrying is Natasha's attitude to anyone who might suffer from profuse, watery diarrhea during the initial stages. In the GAPS book, she says simply to exclude vegetables should it occur, a concerning attitude to a condition that can easily cause serious medical harm, especially in young children!

As the diet progresses, more foods are introduced, but it is the introductory stage, said by Natasha to be essential for healing, that is the most worrying. There is no evidence of this diet having any clinical effect, yet it is being sold as a universal panacea for numerous real, debilitating conditions. As we've seen, the GAPS diet has great potential to harm vulnerable children. But if that was not enough, the rabbit hole goes down much deeper.

THE GAPS RABBIT HOLE

GAPS is big business. Supplements and probiotics form an important part of the program, and although Natasha takes pains to say that she does not endorse any particular brands, she does sell several through her website and regularly recommends the Bio-Kult brand as something followers should introduce. More than anything, the GAPS diet is spread by an army of evangelical followers, many of whom have trained as practitioners on one of Natasha's endorsed training programs. GAPS is no longer the work of a lone maverick in a one-off UK clinic. There are hundreds of practitioners operating around the world. At the last count, there were 346 in the United States, 97 in the UK, and a total of 662 around the world, all selling a restrictive diet with the potential to do great harm.

A closer look at some of the attitudes and opinions of the diet's founder reveals more to worry about. Often citing the work of discredited former doctor Andrew Wakefield, struck off the UK medical register in 2010 for deliberate falsification of research, Natasha has strong antivaccine views, believing that all vaccines can cause GAPS-related conditions to occur. She says that vaccinating against infectious disease causes "an enormous strain on an already compromised immune system, becomes that last straw which breaks the camel's back and brings on the beginning of autism, asthma, eczema, diabetes, etc." To clarify her position, she also comments that:

> Strictly speaking the only two vaccinations that can be considered important are tetanus and polio. Other vaccinations

are not essential; in fact it is better to let your child go through those childhood infections. Just make sure that your child is well nourished and he or she will sail through those infections and come out stronger with a more robust immune system.

Looking at current US vaccination schedules for children, presumably the infections you should allow your child to "sail through" are meningitis C, meningitis B, pneumococcal infections, measles, mumps, rubella, influenza, whooping cough, and *Haemophilus influenzae* type b, which I am not sure many other qualified medical doctors would recommend. Perhaps even more puzzling is her belief in homeopathy. Apparently, "homeopathy combines with GAPS very well and I recommend it. Homeopathic remedies can help you to get through die-off easier and to overcome many stubborn problems. . . . Homeopathy is a wonderful method of healing and can be very effective. But it is a large and complex science and requires serious training." Presumably here she is confusing the word *complex* with *made-up*, as homeopathy is basically the distribution of water and sugar pills.

The concept of "die-off " is often repeated throughout her discussion of the diet's phases, used to explain away weight loss, bloating, skin conditions, vomiting, and pretty much any other symptoms that you might reasonably expect when making a dramatic shift to a nutritionally inadequate diet. We are told that unpleasant side effects are the body's way of removing toxins, and that you should push on through so the healing can begin. Again, I remind you that none of the healing effects are proven, and often this diet is being recommended for vulnerable children younger than three years old.

Other areas where she holds beliefs that would not generally be compatible with a medical doctor are: oil pulling (I wonder if she likes coconut oil); the alkaline ash hypothesis; and, of course, detox, which forms an important part of her program. Natasha is very concerned about the toxicity of modern life, finding sources of dangerous toxins everywhere. She claims that all processed food is toxic and to be avoided, as are swimming pools, cosmetics, beauty products, shampoos, soaps, toothpastes, detergents, floor cleaners, polishes, paints, carpets, furniture, "building materials," washing powders, and deodorants. On the flip side of this, she has a strong belief in the power, wisdom, and general loveliness of Mother Nature. She says, "Man-made things are generally bad for your health and destructive to all life on the planet. Things made by Mother Nature are well-balanced and generally healthy and healing." Clearly she has never heard of botulism toxin or deadly nightshade.

Also of concern is her belief in the healing power of enemas, which she advocates for babies and very young children, a practice that I have to admit I find very troubling.* Enemas are of no known benefit to general or digestive health, and although it might be revealing my own ingrained moral prejudice, I find deeply unsettling the idea of the needless anal insertion of an enema tube into children too young to understand what is going on.

Searching through the website (particularly the comprehensive FAQ section) and the book, there are many deeply worrying statements. The diet advocates feeding raw egg yolks to very young children, against all responsible food safety recommendations. Natasha gives frankly dangerous advice on the preserving of the all-important bone broth by pasteurizing it in jars and storing it for up to a year, risking the growth of dangerous pathogens. She

---

* She says in the FAQ on her website gapsdiet.com: "With babies I recommend to use enemas only for constipation and use only water. In children from around 2.5–3 years of age we can start adding kefir, whey, salt and bicarbonate of soda . . . I know hundreds of families who are doing enemas with their children."

recommends eating liver during pregnancy despite the known risks caused by high levels of vitamin A. She claims that high blood cholesterol is actually a positive sign, showing that the body is healing and detoxifying, at odds with all sensible medical advice. She claims that her highly restrictive diet can cure eating disorders. She even goes as far as to claim that the GAPS diet can alter the human genome, advising someone worried about a child diagnosed with a genetic condition that "the character of toxicity in your child's body will also affect genetics; it will change your genes and their expression."

This rabbit hole goes very deep, and there are many more, equally dangerous examples in the GAPS book and various websites. But according to Brandolini's Law (more commonly known as the Bullshit Asymmetry Principle), the energy required to refute bullshit is many times the energy required to produce it; I only have so much time, and this book can only have so many words.

THE REALITY

I have already received much criticism for my condemnation of GAPS, with many people telling me I am closed-minded, that I should open my eyes and see beyond the need for evidence. But until studies are conducted into the diet's effectiveness, we will never know.

The big question to ask is why those studies have not taken place. Natasha Campbell-McBride has never published a single case study in a reputable journal, despite having treated hundreds of patients. How could this be? Why would someone concerned with the health and well-being of the people she treats not want to prove that her intervention works? Why would someone facing criticism from the medical community not want to prove the naysayers wrong? If the diet really works, it should be studied. It would not be hard to design experiments to test its effectiveness.

Why has Natasha Campbell-McBride never done this? After all, her diet is supposed to cure dozens of chronic illnesses. If she was proved right, she would be written into history as one of the greatest researchers of all time.

I asked Richard Mills from Research Autism about the effect of diet on autism. He told me:

> There is no evidence that any special diets have a role in recovery. Autistic children suffer from gut problems in roughly the same proportions as other children, so sometimes if you withdraw foods and a child has an undiagnosed intolerance of gluten or dairy then there might be some improvement because they are a little happier or not in pain. It is a happy coincidence, but nothing more. If there have never been any properly conducted trials for an intervention, it is impossible to take it seriously.

The reality is that until studies are conducted, no one will ever know what is really going on. There is potentially a double tragedy here, as it might just be that within the records of GAPS treatment, there is some information with the power to help. Zoe Connor, whom I mentioned earlier, is a leading pediatric dietitian specializing in autism and an outspoken critic of GAPS, but is at pains to point out that, for all we know, diet could potentially have an effect (after all, if diet had no effect on health, there would be little need for dietitians). On the large number of supportive anecdotes for GAPS, Zoe says:

> I wouldn't rule out that some aspects of the GAPS diet might "work" for a child with lots of undiagnosed issues. For a child with sub-clinical epilepsy-type brain activity then a low-carbohydrate diet might improve brain function. Also

just the action of pushing a child with autism to make significant changes in eating may have a knock-on effect in making their behavior less rigid in other areas or indeed empower parents to feel able to tackle other problem areas. These are the complexities we are looking at. And this is why dietary changes need to be taken more seriously in the mainstream and studied more. If some parents are saying this diet works, then what is happening? If it is because they have undiagnosed conditions then we can treat that condition using diet or medications and the child can then have a much more varied, adequate, less expensive, and less inconvenient diet with the same improvements.

Never Events

Anyone working in health care will be aware of Never Events. A Never Event is something that should not be tolerated, something that there should be clear frameworks in place to prevent. A child being put at risk of serious nutritional deficiencies is certainly a Never Event, something capable of doing devastating and irreversible harm.

The GAPS diet is sold as a medical intervention to cure specific conditions. There is no evidence that it does so, no published literature supporting it, and nothing but anecdotes reported by its creator. The protocols that a child has to go through if undertaking the GAPS diet carry a real danger of causing lasting physical and psychological damage. This is a Never Event.

Chapter Eighteen

CANCER

*Unlike all the other cells in your body, which can burn carbs or fat
for fuel, cancer cells have lost that metabolic flexibility and can only
thrive if there [sic] enough sugar present. . . . Make no mistake about
it, the FIRST thing you want to do if you want to avoid or treat cancer
if you have insulin or leptin resistance (which 85 percent of people do)
is to cut out all forms of sugar/fructose and grain carbs from your diet.*

—DR. MERCOLA

*A hospice patient admitted for end-of-life care said that one thing she
really wanted before the end was to eat a few slices of French baguette
with butter, but refused the offer from the hospice chef as she was
fearful of taking carbs that would "'feed her cancer cells."*

—CATHERINE COLLINS, RD, FELLOW OF
THE BRITISH DIETETIC ASSOCIATION

It is easy to laugh at the ridiculousness of health bloggers and
wellness warriors, to poke fun at their pretentions and misunderstandings, to smirk at their absurd pseudoscientific beliefs.
Most of the current crop of health bloggers purport only to cure
vague self-reported maladies, such as feeling a bit tired and having disappointing skin, or to help in the achievement of poorly
defined goals like "getting the glow" or "becoming a better you."

However, in this section of the book we have started to glimpse
a world of damaging extremes, where untested treatments are

hailed as cures for real, brutal, and damaging afflictions. Here pseudoscience's dark heart is revealed, and never is the potential price of bad information higher than when it comes to cancer.

Despite the overblown curative claims that Natasha Campbell-McBride makes for her diet, even she steers away from any talk of miraculous cancer cures for her GAPS protocols. She believes that cancer is caused by parasitic organisms that feed on a buildup of toxic sludge in the body. Apparently, when tumors are removed by surgery, tiny parasite babies are released into the bloodstream (her words), explaining why cancers often reoccur. For this reason, she does not recommend GAPS as a treatment: "GAPS Nutritional Protocol is a 'feeding' nourishing protocol. Yes, it provides detoxification too, but in a balance with nourishment. That is why I don't use it in treating cancer." She also helpfully explains that "the time-proven anti-cancer nutritional treatment is the Gerson Protocol, which was developed in the 1930s by a German doctor Max Gerson."

We shall return to Gerson therapy shortly, as it is one of the most persistent of the many supposed dietary cures for cancer, but first it is worth looking at the disease more generally. In diving into its history, some of the reasons for the extraordinary claims and misunderstandings are revealed.

A UNIVERSAL CURE?

Cancer is perhaps the embodiment of our deepest fears. An enemy from within, dark, mysterious, and seemingly unstoppable. It is the killer that hides in our very own cells, consuming its victims with a slow but uncontrollable advance. Growth and cell division, the very things that give us life and vitality, are sent to destroy us, a cruel twist that taunts the modern world. Cancer claims more and more lives each year, despite the advances of science and medicine, and it is now estimated that it will affect one in two men and one in three women at some point in their lives.

As our understanding of health and disease has progressed and many common killers have been wiped from the face of the earth, cancer has remained, dogged, unstoppable, devastating. Each death provides a grim and unrepentant show, played out in full view of loved ones, helplessly watching on as brave souls are devoured from within. Despite remarkable strides in treatment and cures, cancer remains a persistent stain, claiming the lives of an ever-increasing proportion of the population. All of us will brush close to it at some point, and its vile grip will doubtless take the lives of us or of those we love.

Cancer is thought to occur when a single cell somewhere in the body mutates, altering the way it responds to signaling. The circuits that normally regulate cell division and death become broken, producing a cell that will not stop growing and dividing. Although this seemingly simple mechanism lies at the heart of all cancers, that is where the commonality ends. The more cancer has been investigated, the clearer it has become that it is not just one disease. Even specific types of cancer cannot be categorized as a single ailment. Each type is unique, and in every individual case the cancer that presents is singular. In addition, as cancer cells grow with such rapid and uncontrolled voracity they have a tendency to mutate and change, meaning that different cancers can exist within the same expression of the disease. Mutations can occur midtreatment, causing a cancer to develop immunity to different drugs, cruelly morphing into a more virulent disease within the body of the patient. Cancer is a complex, shapeshifting disease that has staggered and confounded the best researchers for a hundred years. Although huge strides have been made in treatment and prevention, the more cancer is studied, the more the prospect of a universal cure seems to vanish on the horizon.

A universal cure for cancer would write a researcher into the annals of history. In 1900, the most common causes of death in the United States were tuberculosis, pneumonia, diarrhea, and gastroenteritis. By 1950, incidence of most of these diseases had been slashed by modern sanitation and medicines. The discovery of penicillin had shown the potential for groundbreaking treatments to revolutionize health outcomes. Vaccines for many common and devastating diseases such as polio had been developed, and revolutionary new drug treatments were appearing at an astonishing rate. In developed societies throughout the 1950s and 1960s, an expectation developed of a future free from disease. It was a world of progress and hope, with many believing that technology and research would defeat all the evils that nature could throw at humanity. But cancer refused to play along, stubbornly resisting any efforts to create a universal cure, the "penicillin for cancer" that the new age seemed to promise. As other diseases were proudly conquered and controlled, cancer climbed ever upward on the list of common causes of death. Increased longevity, superior testing, and a distinct drop in other causes of mortality swiftly revealed cancer's previously hidden power to harm.

A universal cure for this increasingly devastating disease became an obsession, frequently fictionalized, and thought by many leading scientists of the day to be an achievable dream. The Manhattan Project, the development of atomic weapons during the Second World War, had shown the potential for directed science projects to achieve seemingly impossible goals, undermining the previous belief that scientific research should be the preserve of academics driven only by a thirst for knowledge. As public fear spread, there was a call for a "Manhattan Project for cancer" to finally rid the world of this pernicious threat. Inspired by huge progress in the development of chemotherapy, a war on cancer was declared, perhaps mistakenly categorizing it as a single and easily defined enemy. In 1969, the American National Cancer Institute published a full-page advertisement in the *Washington Post* imploring then-president Richard Nixon to act: "Mr. Nixon: You can cure cancer," the advertisement said. "We are so close to a cure . . . we lack only the will and the kind of money and comprehensive planning that went into putting a man on the moon."

The stage was set, and over the next few years Nixon authorized hundreds of millions of dollars to be spent, keen to ensure that the universal cure formed part of his legacy as president. Despite this optimism, and despite vast leaps in understanding, the promised panacea has remained out of reach. But ever since the promises of the 1960s, the world has dreamed of it, even as research has shown that it is likely to be unattainable. As we shall see, the power of that dream has at times led to some dangerous delusions.

A DISTORTED VERSION OF OUR NORMAL SELVES

Perhaps the greatest difficulty faced in battling cancer is that cancerous cells are essentially our own cells, slightly altered. In a speech made by the former director of the National Cancer Institute Harold Varmus on accepting his 1989 Nobel Prize for groundbreaking work on the origin of retroviral oncogenes, he described the battle that researchers face: "In our adventures we have only seen our monster more clearly and described his scales and fangs in new ways—ways that reveal a cancer cell to be . . . a distorted version of our normal selves."

An earlier researcher, William Woglom, described the difficulty in finding treatments that only affect cancer cells but leave normally functioning cells intact as "almost—not quite, but almost—as hard as finding some agent that will dissolve away the left ear, say, yet leave the right ear unharmed."

For this reason, many of the most effective treatments for cancer developed since the 1950s have been profoundly toxic, designed to target the one key feature that distinguishes cancer cells—their rapid, uncontrolled proliferation. Of course, other cells need to divide and grow as well, so any agent that attacks this aspect of cancer will also have a profound and detrimental effect on normal functioning. Chemotherapy regimens, usually a combination of a number of agents that target cell division, tend to be devastatingly tough, producing profound sickness and great damage to the rest of the body so as to destroy cancer cells. Chemotherapeutic agents are toxins, and in the worst cases patients are deliberately poisoned toward the brink of death to destroy the malignancy inside.

Recently a number of more specific anticancer agents have been found, concentrating on some of the other properties of cancer cells, most notably the breaking of signaling pathways and the targeting of oncogenes. These treatments are frustratingly hard to find and usually highly specific, often only having an effect in a small subset of one type of the disease. The work is slow, expensive, and difficult, but it may perhaps lead to a point in the future where many common cancers will be manageable, with treatments that will keep them in check and allow sufferers to lead normal lives. Caution is needed, however, as when it comes to cancer treatment, the approach of many false dawns has been hailed in the past.

THE CAUSE

The complexity of cancer is not just limited to its method of destruction. It also seems that there is no commonly known single cause for cancer-forming mutations to occur. A cancer is thought to start from a single change to a single cell, but it appears that many different factors can influence this mutation to occur. Environmental influences can play a part, with exposure to tobacco smoke, ionizing radiation, sunlight, and inflammation from asbestos dust all known to increase risk. There are a few known viral methods of transmission (such as HPV), genetic factors, and a host of other lifestyle-related reasons that increase the chance of developing cancer.

Diet has a known influence, and there is a strong correlation between many areas of dietary health and cancer risk. For example, the increased risk of colorectal cancer (starting in the colon or rectum) from processed meat consumption, which we discussed previously. But in any discussion of risk it is important to remember that cancer can often occur as the result of seemingly spontaneous mutations. In a lifetime, billions of cell-division events will occur, and it only takes one cell to go wrong in a particular way to devastate a life. There are ways of increasing and decreasing risk, but cancer could happen to any of us. It is the result of a single change, a fleeting and random event, not the long-term degradation of a body nor the direct consequence of specific life choices. The risk of developing cancer is like the risk of any accident. We can avoid taking risks, but we can never outrun chance. Cancer is an unavoidable hazard built into our lives, with perhaps the greatest risk factor of all coming from our tendency to live longer, with greater freedom from other potential causes of demise. When someone develops this awful disease, in any of its myriad forms, the last thing that we should do is blame them. They are a victim of a brutal fate, and we just need to be thankful that in this modern age they have more of a fighting chance than at any other time in history.

THE TEMPTATION OF THE UNIVERSAL CURE

The strides that have been made against cancer—the huge increases in survival rates and long-term outcomes—are largely due to the variety of tools available to modern oncologists, making it possible to treat each cancer through a unique approach.

Treatments are often harsh, punishing, and lengthy, requiring strength, bravery, trust, and determination. Given the terrible battle stretching out in front of anyone at the point of diagnosis, it is not surprising that some people might reach for any branch that promises to pull them in from the stream.

Earlier in the book we discussed the alkaline diet claims about "curing" cancer and misunderstandings about antioxidants, but there are many more. The myths surrounding diet and cancer are myriad, but one of the biggest myths is that there is a single cure or cause for all forms of the disease. This is common for much pseudoscience, but never more damaging than with cancer, where this tempting vision is so much at odds with the real scientific understanding.

Many diet protocols have promised a universal cure over the years, often with roots in the period in history when we still thought it was on the horizon.

Gerson therapy

German-born Dr. Max Gerson developed his treatments for cancer between the 1920s and 1950s based on the supposed detoxification of the body by a specially developed diet and a misplaced belief in the power of coffee enemas. Gerson believed that all disease was a result of imbalances between sodium and potassium and that a prolonged exposure to his extreme regimen would address this imbalance, plus cleanse the liver. This would rid the body of toxins and allow it to heal itself from many different diseases, including cancer. Max Gerson believed that "cancer is not a single cellular problem; it is an accumulation of numerous damaging factors combined in deteriorating the whole metabolism, after the liver has been progressively impaired in its functions."

Even in 1950 this view was a long way from the medical consensus, and as knowledge has developed in the intervening years, it has proved profoundly mistaken. Even so, it is highly persistent with centers in Mexico, Hungary, and the United States still offering the hugely expensive "Gerson therapy," based on the consumption of enormous quantities of pressed vegetable juices, vitamin supplements, injections of animal liver, three to four enemas a day, and rectal hydrogen peroxide treatments. Patients are directed to follow this protocol for long periods of time after returning home, often for years. Gerson health centers distribute literature warning of the possibility of brutal side effects, including chronic pain within tumor masses, fever, cramping, diarrhea, vomiting, and weakness, explaining that this is just the body ridding itself of "toxins." These side effects are unsurprising given the incredibly low sodium levels of the diet and the potentially life-threatening nutritional deficiencies that those undergoing it are subjected to.

Despite its persistence throughout seventy years of medical advancements, and claims that it has "cured" many thousands of patients, no study exists in the literature that supports the clinical effectiveness of the Gerson diet. Max Gerson's book *A Cancer Therapy: Results of Fifty Cases*, published in 1958, has been investigated by the American National Cancer Institute (NCI) and found to contain no evidence of any benefit. When ten "cured" patients selected by Gerson himself were studied by the NCI in 1947, no real improvements caused by his treatments were found. A study of the records of eighty-six patients around the same time also showed no effect. To make things worse, the futility of the treatment is only half the story. In the 1980s a number of Gerson clinic patients were admitted to a hospital suffering from infections received as a result of liver injections, with five patients falling into a coma due to low sodium levels. Over the years, excessive use of coffee enemas has also been linked to a number of deaths from electrolyte imbalance.

Despite no evidence of effectiveness and huge physiological risks, Gerson therapy lives on, risking the lives of cancer patients around the world. Charlotte Gerson-Strauss, the daughter of Max, continues to promote it as effective and has claimed that conventional cancer treatments are harmful. Although many patients come to Gerson in the last stages of incurable disease, the greatest danger of all is that some might be persuaded to reject effective conventional treatments in favor of this seemingly more natural and benign alternative.

When Gerson started developing his treatments in the 1920s, conventional options were few and rife with risk, and the eventual outcome for patients choosing either path may well have been the same. In the modern age, however, when survival and recovery is the most likely outcome for many cancers, to persuade anyone to reject modern medicine in favor of juices, vitamins, and enemas seems at best a pointless risk.

The macrobiotic diet

In 1965 US writer William Dufty translated the writings of George Ohsawa in a book called *You Are All Sanpaku*.* In it the details of "The Macrobiotics" were outlined, a strange quasi-religious movement with the "Zen" macrobiotic diet at its heart. As a result, Ohsawa's teachings took hold within a number of new age communities in the United States in the late 1960s advocating for a largely vegan diet, organic and locally grown produce, the importance of preparation of food in a calm and peaceful environment, and the long, slow chewing of all foods. Its popularity was helped by Dufty's friendship with John Lennon and Yoko Ono and their

---

* Sanpaku *is a term from the Chinese practice of face reading and refers to eyes in which the white space above or below the iris is visible. It is believed that people with the condition are likely to lead violent or accident-prone lives.*

interest in its strange philosophies and perceived Eastern mysticism. Far from being holistic and healing, the Zen macrobiotic diet is brutally restrictive, and Ohsawa made a number of absurd and explicit claims as to its ability to cure all types of cancer. In fact, he went as far as claiming that cancer was one of the easiest diseases to cure, actively encouraging patients to reject conventional treatment. In a 1971 American Medical Association report, the Zen macrobiotic diet was linked to a number of deaths and serious cases of malnutrition.[1] The report condemned it in the strongest possible terms, calling it a "threat to human health."

Subsequent publicity meant that its influence waned, but it resurfaced again in the 1980s, this time promoted by former Ohsawa disciple Michio Kushi. Perhaps inspired by the money machine of Gerson cancer treatment, Kushi explicitly recommended that followers abandon "violent and artificial" conventional treatments and heal themselves using his diet, a program that, if followed in its entirety, could cost hundreds of thousands of dollars. Despite the existence of many testimonials, no published study exists to back up the wild assertions of Ohsawa, Kushi, and the macrobiotic diet. Several requests were made by the American Cancer Society, but Kushi never managed to provide any clinical data to support his claims. Kushi died in 2014 of pancreatic cancer.

Kelley's therapy

Around the same time as Dufty was spreading the word about macrobiotic cures, orthodontist William Kelley was developing protocols based on his theory that cancer was a single disease caused by a lack of certain enzymes and the buildup of toxins in the body. Strongly influenced by the teachings of Max Gerson, Kelley's treatment involved some brutal "detoxification" combined with various other new age nonsense. Steve McQueen was Kelley's most famous patient, approaching him in 1980 when suffering from a malignant and inoperable form of multiorgan cancer called mesothelioma. Later that year, reports spread that McQueen's cancer was in remission after treatment at Kelley's Mexican clinic, earning it a great deal of publicity and interest around the world. But tragically, and perhaps not surprisingly given that the treatment consisted of little more than vitamin supplements, coffee enemas, and prayer (literally, not figuratively part of the treatment), the reports of remission were false, and McQueen died a few months later. The reality of "Kelley's treatment" was revealed in a 2010 retrospective study showing the effects of the diet on cases of inoperable pancreatic cancer. Patients who underwent the treatment suffered a threefold decrease in life expectancy and a greatly diminished quality of life when compared to patients undergoing standard chemotherapy.[2]

There are many more. From Contreras metabolic therapy, Issel's whole body therapy and, of course, the alkaline diet, there are countless people claiming to have knowledge of universal mechanisms and treatments for this terrible disease. The promise of a simple diet-related cancer cure seems almost too tempting for desperate victims to ignore. In a 1993 report, the American Cancer Society concluded:

> Although dietary measures may be helpful in preventing certain cancers, there is no scientific evidence that any nutritionally related regimen is appropriate as a primary treatment for cancer . . . Some involve potentially toxic doses of vitamins and/or other substances. Some are quite expensive. All pose the risk that patients who use them will abandon effective treatment. The American Cancer Society therefore recommends that "nutritional cancer cures" be avoided.[3]

Numerous cancer charities and campaigners have publicly drawn similar conclusions, but still the myths persist, alternative treatment clinics thrive, and victims continue to pay the ultimate price for abandoning potentially life-saving treatments. Misunderstandings about the relationship between cancer and diet continue, and as they do they create boundless suffering.

MODERN CANCER MYTHS

In recent times new myths have developed. Many of these are more related to prevention and treatment of cancers than to claims of outright cure, but they are potentially damaging all the same. The alkaline diet has given rise to many claims that meat and dairy are the causes of a modern cancer epidemic. Unspecified toxins within processed foods are routinely blamed, as is a general pervasive toxicity of modern life. Perhaps the most ubiquitous of all these new fears, however, is the misunderstanding outlined in the opening quotes of this chapter, the belief that sugar feeds cancer cells. As we have seen, sugar has become routinely demonized as a vile and toxic poison, worse than cocaine, so it is perhaps not surprising that cancer has been added to its list of misdemeanors.

In *The Cancer Revolution* by Patricia Peat, the naturopath Xandria Williams claims that any carbohydrate-containing foods, including brown rice and whole wheat bread, are "dangerous foods for you if you have cancer." She goes on to discuss the importance of the Warburg effect, first discovered by physiologist Otto Warburg in the 1920s, which according to Xandria demonstrated that "the prime cause of cancer [is] the replacement of the respiration of oxygen in normal body cells by a fermentation of sugar." Xandria concludes from this that "glucose feeds cancer" and you should "keep that firmly in mind when choosing your foods."

The idea that glucose feeds cancer cells has become increasingly prevalent in recent years, encouraging many cancer patients to engage in needlessly restrictive diets. As with much pseudoscience, there is a grain of truth in the narrative. Unfortunately, it is based on a misunderstanding of the complex biology of cancer cells and quite a lot of wishful thinking. Warburg did indeed notice differences in the metabolism of cancer cells and received a Nobel Prize for his efforts, but it is quite a leap to take a piece of ninety-year-old research on metabolism and turn it into a therapeutic diet regime.

Cancer cells are simply versions of human cells that do not respond to messages telling them when to stop growing and dividing. Like all human cells, they require glucose as a source of fuel. They do tend to have a more voracious appetite than the normal cells surrounding them, which allows them to be detected in a PET scan, but the reason for this is that they are rapidly dividing, a process consuming large amounts of energy. As Warburg correctly observed, cancer cells often use a less efficient method of glucose metabolism than surrounding cells, making the difference in consumption more pronounced.

Most human cells have two ways of converting glucose into energy. The primary method is called oxidative phosphorylation and uses oxygen. When oxygen is in short supply, cells will often switch to another pathway known as glycolysis, producing lactic acid as a by-product. During strenuous exercise, when it is hard to get enough oxygen to our tissues, many cells switch to glycolysis, resulting in the lactic buildup that causes muscle fatigue. Otto Warburg discovered in 1924 that cancer cells often favor glycolysis, even when oxygen is not in short supply. Clearly excited by his discovery (and in need of a bit of a tutorial on hares and lapwings), Warburg postulated that cancer was actually caused by this change in metabolism.

Although the effect is clearly interesting, it is no longer thought to be caused by mitochondrial damage as Warburg postulated.[4] Crucially, it does not really show a distinctive difference between cancer cells and normal cells, just the incredible capacity of all cells to adapt to their conditions. It is now thought that an enzyme called Tumor M2-PK gives rise to the Warburg effect, and although it is not found in the majority of healthy cells, it is produced when rapid cell division is required, for instance during the healing of wounds.[5] The Warburg effect is one of the results, not the causes, of cancer, and although it is rare in normal human cells, it can be turned on at times when fast replication is required. Cancer cells hijack these existing mechanisms to grow with such voracity, still just "distorted version of our normal selves."

The reality of why cancer cells use glycolysis is complex and poorly understood. Most of our understanding of metabolism comes from the study of nonproliferating cells, and as knowledge of rapid cell growth increases, it seems that glycolysis may frequently be preferred even in high-oxygen environments. Glycolysis may be less efficient for energy production, but in dividing cells glucose is not just used for energy—it is also used as a building block to create other cell components, such as fats and amino acids. Perhaps under these circumstances glycolysis is the more efficient system, explaining why it is favored by cancer cells.

There is much uncertainty and ongoing research into this area, but one thing is certain. Starving the body of glucose does not stop cancer growth. To maintain normal functioning, our bodies carefully maintain blood glucose at the same level, synthesizing it from other nutrients if carbohydrate is in short supply. Changing dietary glucose or carbohydrate intake will not have any effect on the amount that reaches cancer cells. Cancer researcher Dr. David Robert Grimes explains:

Like all cells in the human body, cancer cells consume glucose. But the idea you can starve cancer by eliminating sugar is wholly misguided—for starters, cancer isn't picky about where it gets its glucose from; all carbohydrates are broken down into glucose, which our cells need to survive. Simply eliminating sugary foods will not circumvent this, nor will it cure cancer. I suspect the reason this myth is so persistent is that it emits a faint glint of truth, but one that's profoundly distorted—it's good dietary advice to limit one's intake of sugary snacks and drinks, but this good advice is mangled in the telling to something deeply wrong-headed.

A diet low in carbohydrates is likely to place the body under stress and inhibit our natural repair mechanisms at a time when it needs them the most. The only sure-fire way to starve cancer cells is to starve all your cells, a deeply unwise choice when undergoing cancer treatment. Chemotherapy tends to be an unpleasant and brutal affair, and to deprive the body of essential fuel when it is under attack is fraught with danger. Dietitian Catherine Collins outlines the best strategy for coping:

In terms of diet for cancer patients, the core treatment is keeping the background body well whilst the cancer is being treated. It is about optimizing health, the immune system and the nutritional delivery to the other cells. Extreme low-carb diets are not a good way to achieve this and as blood sugar is maintained whatever the diet taken, limiting carbs is of no real benefit. It's been proven that a varied Mediterranean-style diet provides the best all-round nutrition support both in preventing cancer and in reducing the risk of its return. Often patients are struggling to eat and you need to be increasing variety and options, not restricting foods.

Similarly, Dr. Emma Smith from Cancer Research UK, who has written extensively on this subject, says:

> The myth that sugar feeds cancer cells is an unhelpful oversimplification of some complex science. Cancer cells do make energy in a different way to normal cells using lots of glucose, but they are highly resourceful and have even been shown to adapt to feeding on fats and proteins. Research into the way cancer cells get and use energy is an important area of study and could lead to treatments that target these metabolic processes.

Also worth noting is that the Warburg effect, although quite prevalent, does not occur equally in all cancers. Certain specific tumors show it to a large extent, but others less so. In somewhere between 10 and 40 percent of cases, the Warburg effect does not occur at all,[6] so it certainly is not the ubiquitous mechanism for cancer progression that many claim. As with much pseudoscience, when it comes to sugar and the Warburg effect, a leap has been made from hypothesis to treatment without taking the crucial step of collecting clinical evidence. The Warburg effect is real, but sugar feeds all cells, not just cancerous ones. Cancer is a distorted version of us, feeding and growing in much the same way, which is exactly why it is so hard to fight.

THE PAIN OF FALSE HOPE

Although it is hard to determine the true cost of alternative cancer treatments, it is clear that false hope is a tragedy. In 2008, at the age of just twenty-two, Jess Ainscough received a particularly awful diagnosis. Some strange lumps had been popping up on her left arm and a biopsy revealed them to be caused by an epithelioid sarcoma. This is a particularly rare form of cancer, one that develops slowly and is unresponsive to most chemotherapy or radiation treatments. An initial attempt at isolated limb perfusion, a process in which her arm was isolated from the rest of her circulation and subjected to high-dose chemotherapy, was initially promising, but tragically the cancer returned a year later. The only conventional treatment option available to Jess was surgery to amputate her left arm and shoulder. Even with the bleak prospect of this dramatic and disfiguring operation, there was still a chance that the cancer would reoccur after amputation. For a young woman, who just a short time previously had the world at her feet, the thought of such a seemingly medieval treatment being her only option must have been almost inconceivable. She would be brutally damaged and still face the prospect of the monster at the door. It must have shattered her world.

It appears that somewhere along this terrible journey, Jess started to blame herself. She described her life before diagnosis, based in Sydney and working as the online editor of *Dolly* magazine, as "burning the candle at both ends, paying no attention to how my actions could affect my health." In subsequent articles she is quoted as saying, "I was living a life of excess and disconnecting myself from nature, [which] eventually led to my body manifesting cancer." When the only treatment option offered was devastating surgery, she decided to turn her back on conventional medicine. In her words, she decided to

> take responsibility for my illness and bring my body to optimum health so that it can heal itself. For me it was an easy decision. I began looking at the different ways I may have contributed to the manifestation of my disease and then stopped doing them. I swapped a lifestyle of late nights, cocktails and Lean Cuisines for carrot juice, coffee enemas and meditation and became an active participant in my treatment.

Clearly at some point she had fallen under the belief that some aspects of her lifestyle were the cause of her disease, and was searching for salvation, looking to somehow cleanse herself back to health. In a deeply unfortunate twist of fate that would lead to a double tragedy, Jess had stumbled across Gerson therapy.

As she embarked on what she believed was a healing journey, Jess started to write about her experiences as "The Wellness Warrior." Jess was young, photogenic, and extremely likeable, and her story captured the hearts of many in Australia, propelling her into the media limelight as her story progressed in apparently miraculous fashion. It is impossible to read her blogs or watch her videos without being struck by her remarkable courage and positivity under the most tragic of circumstances. The Wellness Warrior brand flourished, quickly establishing Jess as one of Australia's top health and well-being celebrities, with books, extensive media coverage, and merchandising sales turning her journey with cancer into a powerful money-making machine. Although she denied it in later years, Jess frequently claimed that she had cured her cancer, purporting to be a survivor who had fought off her disease by natural means.

She visited the Gerson clinic in Mexico in 2010 and became a strong advocate for the therapy, explaining the importance of attending one of the enormously expensive Gerson clinics to heal. Talking about Gerson therapy in an online video, she says that "you can't take shortcuts. It's all been scientifically worked out. If you do anything away from it, it won't work. . . . It doesn't heal selectively; it heals all cancer."

Jess certainly did not take shortcuts. Continuing with the therapy at home meant that every day for two years she consumed twelve glasses of vegetable juice, performed four coffee enemas, took staggering amounts of supplements, and existed on a highly deficient diet. Perhaps most tragically of all, in the midst of this punishing regime, she wrote, "The one thing that is pushing me through the therapy is the thought of how amazing it will be once I finish."

All this time, Jess's cancer was slowly progressing. Even though she saw signs of improvement and felt inspired that her natural methods of treatment were having an effect, the disease was advancing in exactly the same way that it would have been expected to do so if untreated. Jess was diagnosed in 2008 and, without treatment, the average survival time for her form of the disease is between five and ten years.

In 2011 the story took a dramatic and ultimately devastating turn. Jess's mother, Sharyn, thought to have strongly influenced Jess's decision to "heal naturally," was diagnosed with breast cancer. Inspired by her daughter's apparent success with Gerson therapy, Sharyn rejected all conventional treatments, refusing a mammogram, biopsy, surgery, chemotherapy, or radiotherapy, instead putting her faith in Gerson with its coffee enemas and juice diets. Sharyn survived for two and a half years, which is about the average for untreated breast cancer patients, before succumbing to the disease late in 2013. Had she not refused conventional treatments, the likelihood is that she would be alive today.

Devastated by the loss of her mother, Jess largely retreated from public life, but to her many followers something else seemed to be dramatically wrong. In rare photographs and public appearances she seemed to be hiding her left arm, leading to speculation that her condition was deteriorating. In 2014 she revealed that things had indeed taken a turn for the worse, and a tumor mass in her shoulder had eroded through her skin. There were many reports that she returned to conventional treatments around this time, perhaps in a last-ditch attempt to prolong her life, but in reality, from the moment she rejected conventional medicine, there was little hope. In February 2015, seven years after diagnosis, Jess Ainscough tragically died, aged just thirty. Although her chances of long-term survival were never high, with conventional treatment the average ten-year survival rate for patients of Jess's age with epithelioid sarcoma is around 72 percent. She would have had a chance, and most likely would be alive today.

What I feel most about the Jess Ainscough story is a sense of waste. Her struggle was unimaginably hard, and the regimes she subjected herself to in the last years of her life were brutal, but that pain seems so much greater because of its futility. Jess had bravery, fortitude, and a strong will to live, but she was handed the wrong tools for the battle. If she had fought with the help of conventional medicine, maybe she would have had an even more powerful story to tell.

Most of all I am angry that she blamed herself for her disease. By all accounts she had led a perfectly normal life until fate dealt her a life-changing hand. She should never have been made to feel like her disease was a punishment for some perceived transgressions. Like all disease, cancer does not make moral judgments. It is a cruel and random affliction, but one whose cruelty modern medicine is slowly winning the battle against.

THE TRUE COST OF FALSE BELIEFS

It is impossible to calculate the cost of the myriad false beliefs surrounding links between diet and cancer. Many sufferers will embrace alternative treatments alongside conventional ones, often wrongly attributing cures or improvements to dietary changes, when in reality it is the proven medicines that have the effect. The website What's the Harm? describes many victims of alternative medicine, including poor souls like Sharyn Ainscough, but in reality it is often hard to tell what the outcomes would have been if people had chosen a different path. Everyone has the right to choose, but anyone who does seek alternative diet-based treatments for a disease as serious as cancer should have the words of Cancer Research's

Dr. Emma Smith ringing in their ears: "There is no good evidence that any type of special diet can help cure, treat, or prevent cancer."

This is not the biased view of the paid shills of Big Pharma. This is the accepted view of serious independent researchers, built from generations of study and research. Scientists are loath to make unequivocal statements, but in this case it is quite clear. There is no evidence. I am a firm believer in everyone's right to choose, but when all options look torturous and difficult, as is surely the case at the point of cancer diagnosis, we are vulnerable to taking unwise risks.

If you embrace new age beliefs, if you detox, cleanse, heal with crystals, support homeopathy, or use unproven herbal treatments, you need to understand that this is where it leads. If you reject the orthodoxy, dismiss nutritional science as broken, and look to ancient wisdoms for answers, this is also where you are headed. The more pseudoscience is reified by society, the greater the chance that at the point of cancer diagnosis, someone will look at the terrifying options stretching out in front of them and choose a path that will rob them of life. The more we detach ourselves from reason, the more likely it is that another desperate victim will spend their last days undergoing painful, abusive treatments in the vain hope of recovery, bereft of the dignity that their brave struggle deserves.

The end point of that stupid, pointless detox salad you chose for lunch lies here. It is people claiming they can cure deadly disease with carrot juice and enemas. It is clinics extorting the life savings of the dying. Put down that superfood smoothie, ditch your wellness books, close that clean-eating web page and delete it from your history. Unfollow the wellness gurus who encourage you to abandon the tenets of scientific reason. It might seem harmless today, but sometime in the future, when your options are not so bright, it may cost you more than you can afford.

Part V

THE FIGHT BACK

Chapter Nineteen

THE EVOLUTION OF MYTHS

I know I said that in the pages of this book I wouldn't offer the one magical secret of how to eat healthily, but I have come across something that I cannot help but share. I think I have unearthed an interesting piece of science that might just solve a lot of the health problems of the modern age. I would like to introduce you to the next big thing in dietary health trends. It's called the karyo-type diet.

Hurrah! We're going to be rich.

Although it originates from theories developed within the complex science of cytogenetics, the premise is really quite simple. It is based on the principle that the best way to eat is governed by our natural place in the food chain. As a species, humans have rapidly developed technologies that give them superiority over other forms of life, thus moving us beyond our natural positioning. For instance, we have developed different ways of catching many types of fish, yet throughout most of evolution we would not have been able to do this. We have also created guns and traps to hunt or snare a number of other creatures that would not have been available to our ancient ancestors. For this reason, we are regularly eating many species that we are not genetically predisposed to consume.

How can we tell which of earth's creatures might make suitable foods for us? The answer is simple. It lies within our chromosomes, the holders of our genetic material that defines who we are.

Ah! I remember chromosomes from school.

Our natural place in the food chain is clearly defined by the number of chromosomes contained within our cell nucleus. Certain "higher" animals and plants are unsuitable for us because they contain more genetic material and so occupy a place above us in the chain. Eating them interferes with our metabolism at a deep molecular level, and studies have shown that consuming large quantities of products made from these species can cause metabolic problems such as obesity and diabetes.

Other "lower" creatures or plants with small numbers of chromosomes are more suitable, so if we stick to eating only these, our health and well-being will be dramatically improved. The lower the chromosome number, the more appropriate the food. I have seen people lose weight after starting on this radical diet program, as well as improve their general health, develop a natural glow, become full of vitality, and have perfect shiny hair.

To follow this diet, there are a number of foods you should definitely avoid. Remember, humans have 46 chromosomes in each cell nucleus. Certain fish, such as carp (104) and shrimp (86 to 92), should be avoided as these would have been nearly impossible to catch before the advent of such technologies as lines, hooks, and nets (carp are known for their elusive intelligence; shrimp are tiny and hard to snare). Dogs (78), horses (64), and hedgehogs (a huge 90) are also off the menu, clearly unsuitable for human consumption. So are goats (60), sheep (50), red deer (68), hares (48), turkeys (80), pigeons (80), and chickens (78).

Nooo! Not chicken nuggets. I thought they were a superfood.

Most plants are allowed, but there are a few that for complex genetic reasons are also too high in the chain. Potatoes and tobacco (both 48) are both full of harmful chemicals that have origins deep within their chromosomes, explaining why tobacco smoke is so toxic to humans.

Some plants and animals have chromosome numbers close to ours. Badgers (44), rabbits (44), oats (42), and wheat (42) should only be consumed infrequently. Others, such as pigs (38), earthworms (36), and tigers (38), should be eaten only semiregularly.

Mmm, delicious earthworms.

Get below about 25 and you are reaching the point where cytogenic suitability is at its maximum. Rice (24), snails (24), and beans (22) are all highly suitable ingredients, as evidenced by the good health of the people of Valencia, Spain, who regularly consume paella with white beans and snails. Cabbage (18), radish (18), and kangaroo (16) are chromosomal superfoods. Peas (14) and barley (14) are full of essential micronutrients all made possible by their chromosomes. If you can get hold of a female Indian muntjac deer (6), then you will be on the road to perfect health, explaining why many practitioners of Ayurveda consume this nutrient-rich meat. Garnish it with some male jack jumper ants, with an incredible one chromosome, and you will be filled with glowing vitality.

Anyone who has read my blog, or this far into the book, will know how cynical I am about fad diets, but I came to cytogenic eating when I was in a terrible period of ill health. I was constantly plagued by colds, throat infections, and general lethargy. I had terrible bowel problems, a mix of chronic constipation one week and foul-smelling, watery diarrhea the next. Things were so bad that I was going to have to give up writing due to exhaustion and numerous incessant health problems. But as soon as I started to cut high-chromosome foods from my diet and focus on the healthful low-chromosome varieties, gradually my health was restored. I was full of life and energy, allowing me to write these very words. If you or the people you love and care for ever feel tired, confused, sick and out of sync with the world, then switch to cytogenic eating and the karyotype diet. I guarantee that you will lose those excess pounds and find a happy, healthier, and longer-living you.

IDEAS THAT STICK

Obviously the karyotype diet is garbage (junk DNA, to make a geeky molecular biologist joke). I made it up just yesterday morning when I was out running and researched it for a few minutes on Wikipedia (who knew lampreys have 174 chromosomes?). The reason for including it here is to explain some of the factors about myths and fads that help to make them stick in people's minds. I like to think that the karyotype diet fulfills most of the essential criteria and, to be honest, it has just as much scientific validity as many of the fads we have discussed so far. Certainly there is very little in the last few paragraphs that could actually be described as scientifically inaccurate. The nutritional content of foods is largely dictated by the actions of the DNA contained within its chromosomes. Anyone attempting the diet would encounter a strict set of rules that if followed would be highly likely to lead to weight loss. I have even seen a national newspaper article where someone claimed we should not eat octopus because it has more chromosomes than we do, indicating that some people are susceptible to the idea that this number somehow indicates evolutionary complexity (if it did, potatoes would be three times more complex than kangaroos). And although my new diet was created in haste and without any scientific rigor, there are other reasons why I think that the karyotype diet might just have the potential to take hold.

In their book *Made to Stick*, Dan and Chip Heath examine the factors that make it possible for ideas to proliferate and grow.

Although the book is mainly about embracing positive ways of getting messages across, much of it is based on insights into how false ideas proliferate. Chip Heath has spent a great deal of his academic life studying urban myths, investigating how some stories are relayed around the world for many years by word of mouth without a shred of corroborating evidence. These myths cross a wide range of subjects, including stories about gangs of kidney thieves, Kentucky fried rats, murdered teenagers, and anally inserted rodents, but all meet a certain set of criteria to help them spread. In addition, they tend to develop over time, changing to make them more effective in self-proliferation. Just like the evolution of species in the natural world, ideas constantly change, preserving certain properties, but adapting to local conditions, new technologies, and societal change.

Their book outlines six main criteria that need to be met for ideas to "stick." These criteria do not just relate to myths and urban legends; they concern any idea with the potential to take hold, including dietary myths and fad diets. "Sticky" ideas should be:[1]

1. **Simple**—The karyotype diet has a simple premise, the categorization of all foods into good and bad groups based on straightforward, easily researched numbers. Much like Paleo or alkaline, it helps if rules are created based on vaguely scientific principles. It should be possible to communicate any good idea as a simple Hollywood-style movie pitch, delivering the core in just a few words. "Eat according to your chromosomes," "Eat alkaline foods," or "Eat like a cave dweller."

2. **Unexpected**—This diet certainly has a sense of the unexpected about it. The hidden secret to healthy eating has always been locked away within our cell nucleus. The reason

we shouldn't eat potatoes, dogs, and hedgehogs is finally revealed. It also creates a "curiosity gap" where a truth is revealed, but we leave the audience wanting more information. People will want to discover which plant and animal species have hidden eating rules within their chromosomes. Should we eat beef? Tomatoes? Avocados? Mulberries?

3. **Concrete**—Ideas should be concrete in people's minds; emotive language should be used to paint clear pictures. In the karyotype diet, distasteful and unsuitable ingredients are deliberately used, lumping them together with common animals and plants to create unpleasant associations. A few vivid descriptions of bowel movements are also employed, always popular within the health blogosphere for creating associations of disgust.

4. **Credible**—Ideas need to have some credibility. I may need to work on this a little for my diet, but little within the diet description I created is untrue. For instance, it is not technically untrue to say that "studies have shown that consuming large quantities of products made from these species can cause metabolic problems such as obesity and diabetes," since consuming large quantities of most things will lead to those problems. I could quote scientific literature to prove that the nutritional content of foods is directed by the DNA within their chromosomes. It would be relatively easy to obtain a spurious PhD of my own so I can start using Dr. Angry Chef as a more influential moniker. Failing that, I could always fall back on my ancient biochemistry qualifications and my twenty years' experience working with food.

5. **Emotional**—Strong ideas always relate to people, not numbers or concepts. Again, this needs a little work and I will require a few more powerful anecdotes to prove my case, but for now, inspired by many a health blogger, I will rely on my personal journey to add power to the story. It is always good to try to draw on tales of health problems that most people can relate to. Many people worry about their weight—and who hasn't felt tired or suffered occasional digestive problems?

6. **Stories**—Similarly, ideas are easily remembered and will spread if they are related through stories. Anecdotes help, but there should also be a strong narrative running through the idea itself. The karyotype diet is about humans, evolving tools and technologies, and hunting for foods they are not genetically adapted to consume—and so has the potential to create powerful visual images. Vivid language is used to create an idolization of the past and a pervasive fear of modern life.

Although not discussed in the Heath brothers' book, within the world of food and health, there is a final criterion that needs to be fulfilled for something to take hold. To thrive, diets do need to work. This does not mean that they have to fulfill all the wild health claims they make, nor does it mean that the premise underlying them needs to be true, but generally speaking they are most effective if they result in weight loss. In the modern era, weight loss has become strongly associated with health, and so it is easy for people to see the losing of pounds as indicative of many other unseen but harder to measure benefits.

Obviously, my diet idea does lack one crucial factor in its potential proliferation. As Malcolm Gladwell asserts in his book

The Tipping Point, ideas need to be adopted by influential groups and to reach a certain critical mass before they can explode into the mainstream. But even if it never goes global, in the modern information age, there is always the ability to find a niche. Social media creates many closed, self-affirming populations capable of perpetuating and reinforcing falsehoods. Ideas don't have to travel the world to do a lot of damage.

THE EVOLUTION OF BELIEFS

Pseudoscientific beliefs are mostly born out of misunderstandings of science, based on grains of truth that are overextrapolated many times to become vast monsters of unstoppable woo. (*Woo*—a term commonly used within the skeptic community referring to false beliefs that dress themselves in the trappings of science, often using sciencey-sounding terminology.) The concept of detox is ubiquitous and widely believed, yet without any clear origin. It has grown from nothing into a vast bullshit-octopus with tentacles spreading into every facet of life. It has done this against the odds, with thousands of sensible, informed commentators decrying it as nonsense, but it persists, and as it does, it evolves and changes. The same is true of the majority of nutritional pseudoscience. Often it is possible to trace the creators and plot a history, but the most powerful and influential ideas tend to have a life of their own. They are beyond any control, flourishing within the hive mind of social media, strange hydras that can become almost impossible to fight.

Just because something has evolved without conscious thought or planning does not mean that it does not serve a specific purpose. We have two eyes so we can judge distance and survive after injury, but no one ever planned our stereoscopic vision. There was no committee, no list of pros and cons, no inventor, no patent filed. It just happened, the product of a million random genetic accidents, missteps, and blind alleys (perhaps literally), tracing back to the longitudinal symmetry of our ancient aquatic ancestors. Evolution is nothing but a series of random accidents, but it does have an incredible ability to create ingenious solutions. We are the products of randomness, but the results are far from random. Like any form of life, we are sleek, purposeful, and brilliantly adapted to survive.

Dietary fads evolve and proliferate in much the same way. The really effective, convincing bits of pseudoscience are the ones that evolve and grow. This evolution is random and largely undirected. The alkaline diet has evolved immensely since its origins in the early days of nutritional science and even in the short time since the publication of Robert Young's books. It has been forced to rapidly change since his arrest, in an attempt to distance itself from explicit claims of cancer cures. Detox, too, has evolved and grown as companies and individuals cash in on its easily believed ubiquity. Paleo has also evolved, which is ironic given that it is based on a misunderstanding of evolution. Maybe if the karyotype diet was ever released into the world it would do the same, changing and growing to become an accepted truth.

It has been suggested to me that the Angry Chef is a slight misnomer, because I believe that most dietary fads are spread by people who are simply misguided. It is quite rare that my anger is directed at individuals (although that does happen), but there is still plenty to be angry about. Just because dietary fads are not created with pernicious intent does not mean that they are not destructive. Detox creates fear, guilt, and shame but usually does so without the malicious intention of any one individual. As dietary myths evolve, we collectively nurture them, so it is possible to feel rage, especially when you see how destructive these myths can be. Although not conscious beings, they can be blamed, and, crucially, they can be fought. It is not easy though. Fighting a misguided individual is hard, but confronting an entire belief system can seem nearly impossible. At times it can make you feel like a drunk in the parking lot of a bar, raging and swinging at the world.

THE FIGHT

Just as with any process of natural selection, the way to fight evolving and ever-changing myths is to create competition for the limited resources they need to thrive. They are too developed to be cut off at the root, so you must catch them where they feed. Messages of truth must fight hard for the precious and limited resource of people's minds.

To do this, it needs to be possible to make evidence-based messages that appeal in a more powerful way than the dietary myths. But here lies a huge problem. Although real messages of dietary health can fit many of the criteria required to make ideas stick, they are constrained by the truth and will always struggle in some regards. Sensible talk of moderation, small improvements, and slow incremental changes will never create anecdotes as powerful and emotive as those of pseudoscience. It may be easy for science to deliver credibility, but in a world where society's relationship with "mainstream" thought is troubled, and science is often painted as corrupt, even this can be difficult. And although nutritional and food science has revolutionized everyone's lives in the past hundred years, its effects have generally been steady and long term, lacking the sort of powerful and emotional story required to go viral.

It is not enough to tackle dietary myths in isolation, attacking each one with competing evidence-based messages. To sell sensible, truthful messages, scientific truth itself needs to be made into an idea that sticks. In its entirety, the world of science has enough power. It has emotive and powerful stories; it has saved countless lives, produced extraordinary heroes, and explained many strange, unexpected, and wonderful things. Although it is a thing of staggering complexity, at its heart is simplicity, the idea that we never really know—that questioning is what drives us forward. Science needs to be made to stick, a powerful idea that will stop the nutri-babble and lies in their tracks.

Angry Chef's deceptive legacy

If this book is to have any sort of legacy, I hope it is this. I would like to propose a small experiment. Most people in the world will never read these words, but for those who do, I would like you to set about planting seeds of the karyotype diet and cytogenic eating in your interactions with the world. In appropriate conversations, occasionally suggest that the key to perfect health is to eat based on the chromosome number of the species we consume. It is my hope that if this seed is planted in enough places, somewhere it will take hold.

Perhaps a misguided or unscrupulous health blogger will pick up on the ridiculous claims and start to spread the word. I hope that the idea will stick in a few people's minds and that for a brief moment (not enough to do any harm) it is believed and shared. Then, we can expose them for what they are, peddlers of dangerous nonsense, insensitive to evidence and highly susceptible to easy messages. Maybe, for just a few people, that would cement the need for an underlying belief in scientific reason and help stop the bullshit from taking hold.

Chapter Twenty

SCIENCE AND TRUTH

Paltrow Science: OK, you win, Science Columbo. But just answer me this: How did you manage to figure it out?

Science Columbo: Do you know what, Paltrow Science? I had my eye on you from the very beginning. From the very first time I saw you talking about antioxidants. You were just so sure of yourself. In my experience, whenever anyone is so certain, especially about something as complex as antioxidants, it usually means they have got something to hide.

Paltrow Science: But that makes no sense. You are always sure of yourself.

Science Columbo: Oh no. Just ask my wife, I drive her crazy. I don't sleep some nights, always trying to get things straight in my mind. There is always one more thing bothering me.

Paltrow Science: That's no way to be. You should relax more. Maybe watch some of my mindful wellness videos.

Science Columbo: I don't think that's such a good idea. If these details didn't bother me, I'd never get to the truth. And then where would we be?

Paltrow Science: Well, I wouldn't be going to science jail.

Science Columbo: I guess not. Come on. Time to go.

WHEN SCIENCE GETS THINGS RIGHT

Charles Townes changed all of our lives. While at Columbia University in 1951, studying the way that microwave radiation and molecules interact, he realized that it might be possible to create precise beams of radiation. Many of the leading physicists of the time, including Niels Bohr, the founding father of quantum theory, claimed that such a beam was impossible, as it violated Heisenberg's uncertainty principle. Many others complained that it was a pointless exercise, extremely difficult to achieve and with no discernible practical use. For some reason, few people saw any need in their lives for a coherent beam of microwave radiation.

Most of us would have given up in the face of such criticism, especially when it came from esteemed people working in the same field, but Townes was a determined man with a history of achievement in developing radar systems during the Second World War. He was convinced that his concentrated beams of radiation would have an important use in studying the structure of atoms and molecules and decided to persist.

Townes and his team started working on his device, and by 1954 they had proved Bohr wrong by creating the first amplified microwave rays by stimulated emission. Townes also seemed to have a natural talent for marketing, and eager for a catchy acronym, the team came up with the name MASER (microwave amplification by stimulated emission of radiation). A few years later he also showed that theoretically the same thing could be done with visible light, so conceiving the laser.

Charles Townes died in 2015, aged ninety-nine, and achieved much in his life, including leading the Technology Advisory Committee for the Apollo flights, but it is the work that led to the creation of laser technology that we should thank him for the most. Breakthroughs in science are never truly the work of one individual, but there is a fairly good argument that without Charles Townes we would not have lasers. The technology and theoretical knowledge to create them existed in the 1930s, but because of the practical difficulties, the theoretical uncertainties and the seeming futility, no one was interested in working on them.

The development of masers was driven by a desire to understand more about the structure of atoms and molecules. For many of us, anything driven simply by the need to understand can seem fruitless and impractical. What is the point in wasting all that resource on things that will never have any real-world applications? Why bother with academic research? If it is all "academic," it will never have an impact. Most of us do not care about the fundamental structure of molecules and atoms, about the nature of quantum particles or the stimulated emission of radiation. Few of us really understand the point of the Large Hadron Collider and many think that the billions spent on it could be put to a more practical use.

But, thankfully, people like Charles Townes have a burning curiosity to discover, to discount the practical and seek only to further our knowledge. His detractors, especially those who said that his ideas would have no use, could not have been more wrong. Lasers are now essential for modern communication, transporting vast amounts of data through fiber-optic cables. Without them the communications revolution would never have happened. Every day we interact with huge amounts of information delivered to us with incredible clarity and speed, thanks to laser technology. They are also used in surveying, reading bar codes, getting information from storage disks, surgery, engineering, creating holograms, printing, etching, and countless other useful tasks. Just sixty-five years since Townes' experiment, hundreds of millions of lasers now exist, affecting every aspect of our lives, and new uses are constantly being found for this most versatile of technologies. Perhaps even more important, they are also used extensively in the way that Townes imagined, in researching some of the most fundamental questions we have about the nature of matter. Currently a number of ultrapowerful lasers are being developed and used around the world to replicate extreme events in the formations of stars and to investigate the nature of the early universe. These lasers are complex, expensive to build, and seem to have few tangible benefits to everyday life, but for anyone criticizing the nature and expense of this work, it is worth remembering the similar criticism that Townes and his team received in 1951. These lasers are at the forefront of fusion technology and numerous other areas of potentially groundbreaking research.

Science is littered with such examples, a few of which we have already discussed. Niels Bohr might have been wrong about lasers, but his work on the quantum nature of electrons in the early twentieth century had a number of profound and unforeseeable effects many years later. Without it, we would never have had the transistor; without that, there would be no microchips; without them, no computers, no internet, no smartphones, no YouTube, no sharing of funny cat videos.

For an example a little closer to the subject of this book, in the early 1980s Australian researchers Barry Marshall and Robin Warren became convinced from epidemiological studies that *H. pylori* bacteria caused peptic ulcers and gastritis. This claim was ridiculed by the scientific community, who held the consensus view that these were lifestyle diseases and a bacterium could not survive the acidic environment of the stomach. With little support for their ideas, Marshall decided to drink a petri dish full of *H. pylori*, and a few days later developed symptoms including vomiting and gastritis, subsequently curing himself with antibiotics. This experiment went on to revolutionize the treatment of these conditions around the world, relieving the suffering of millions of people.

THE IMPORTANCE OF FAILURE

Throughout this book I have at times criticized the world of nutritional science for many things: its obsession with reductionist experiments that do little to shape the narrative when it comes to making decisions about food, the arrogance and susceptibility to bias of many specialized academics, the inability to provide a consistent voice confusing the public by creating needlessly sensationalistic headlines that overstate the importance of their work. But the fact remains that science produces the greatest work of humanity. For every Charles Townes, there will be thousands of others working just as hard and with just as much dedication, but without the final payoff. There are only so many world-changing discoveries, but those discoveries will never be reached without the activities of the hive mind of science. Countless fruitless experiments are destined to change nothing. The vast majority of scientists will retire without a Nobel Prize or a world-shattering breakthrough to their name. But these "failures" are not failures at all. They are just as responsible for the progress of science as Townes, Bohr, Marshall, or Warren, because they explore all avenues and lead to the successes of others. It is impossible to predict where the next world-changing discovery will occur, so we need to trust the process, because it delivers time after time.

I will always decry anyone who makes wild insinuations, extrapolating the results of test-tube experiments into real-world outcomes, but that does not mean that the experiments they refer to have no value. Alexander Fleming's original observations were in a petri dish and they went on to change the world. Many other in vitro observations will change little, and precious few will change the world, but they will push human knowledge a little further and in doing so edge us ever closer to the next breakthrough.

If you want to really understand how to spot bad science, there is one simple rule that overrides all the others I have given you so far in this book. The simplest way to spot false prophets is that they will always be 100 percent sure of themselves.

A true scientist is not defined by any measure of qualification, cited research, or history of groundbreaking discovery. A true scientist is someone who is aware of the limitations of their own knowledge. They will always doubt, they will always question, they will always be bothered by just one more thing. Anyone who tells you that they know, that they are sure, that they have no doubt—they are either ignorant or dishonest. The moment that doubt stops, science and progress end. Only someone willing to doubt the accepted view will create a laser, conceive a semiconductor, or drink a glass full of bacteria.

THE SHAMING OF EXPERTS

As education has increased and access to information has been democratized around the world, there is an ever-increasing tide of expert shaming and the rejection of consensus views in favor of outliers. Of course, the existence of doubt and disagreement is the engine of scientific progress, but with this democratization comes great danger. The scientific community has always held consensus views, and its disagreements have always taken place out of the view of the public. Today these disagreements take place in the full glare of the world's media, and the views of outliers and fanatics can achieve the same status as those of the whole scientific establishment. Disagreement happens at the front line of science, but there is a rich seam of careful experimentation and evidence that lies beneath, well-tested theories that we can trust. Townes was a pioneer of lasers, but he needed to understand Einstein and Bohr's work to get there. Marshall and Warren challenged many assumptions, but their work can be traced back to Louis Pasteur.

The culture of expert shaming is one of the most pernicious recent developments in our society, with the potential to do great harm. To question science is to ignore everything it has done for man, to overlook the astounding progress of the last few hundred years. The irony of people questioning what science has done for us while typing on a computer, connected to the internet via a fiber-optic cable, should not be lost. The stupidity of doubting the expertise of dietitians and nutrition researchers when we live in a world where the study of such things has transformed life expectancy, quality of life, and health outcomes should always be used to hold the expert shamers accountable.

We need experts. We need them to drill our teeth, to fix our laptops, to build bridges, and perform surgery. We should not pick and choose when we engage with science. We should not decide that in some areas we have the expertise because of some instinctive sense of what is right. Just because the outputs of nutritional science can seem less tangible than those of engineers, surgeons, or computer programmers does not mean that they are any less significant.

Scientists have already discovered ways of processing water to save millions of lives. Food science created preservation and storage techniques allowing nutritious foods to be consumed all year round. It has transformed the food chain, freeing billions of people from hunger and want. It has lessened the incidence and severity of catastrophic famine events that have blighted much of human history. It has discovered vitamins enabling the prevention and treatment of devastating deficiency diseases. It has isolated insulin to allow type 1 diabetics to live beyond childhood. It has developed a simple and effective oral hydration therapy for diarrhea that has saved millions of lives around the world. It has nearly doubled our life expectancy in the last hundred years. It is mad to say that it is broken.

THE BEAUTY OF OUR IGNORANCE

I am not a scientist. Every word I have written in this book can be challenged because of my lack of credentials, my lack of PhD, my lack of published research. I am just a chef torn between two worlds. My love of food has defined my life and my passion for science fills my thoughts. I remember very little of my time spent studying biochemistry, and even if I did the science has moved on so far since then as to be unrecognizable. But it did leave me with a sense of what it is to be a scientist.

How can science continue to change the world in an age where access to information has been completely revolutionized? In many areas this is simple, but when it comes to food, there are great challenges. The more we distrust our experts and reach for the quick answer, the more we leave ourselves open to the fanatics and charlatans. The responsibility lies with normal people like us.

Perhaps one of the greatest scientific endeavors of all time, the Apollo missions to land a man on the moon, was inspired by a speech made by John F. Kennedy in 1961 to a joint session of Congress, where he outlined what many at the time thought an impossible goal. It is a remarkable and inspiring speech in many ways, but perhaps the key passage is this: "In a very real sense, it will not be one man going to the moon—if we make this judgment affirmatively, it will be an entire nation. For all of us must work to put him there."

The inspiration came from imbuing everyone with a sense of responsibility for the outcome and with a sense of pride in the achievements. Landing on the moon was not just the triumph of a few brave men, and it was not the success of a few brilliant scientists. The mission's success depended on a nation, perhaps even the whole world, supporting the endeavor. If there had been doubt, if the world had asked what the point was, if nations had railed against it, doubting the experts and spreading fear, it never would have been achieved.

The following year, Kennedy gave a speech to thousands at the Rice Stadium in Houston, Texas, explaining the reasons why he had committed the nation to such a lofty goal. He said:

> We set sail on this new sea because there is new knowledge to be gained, and new rights to be won, and they must be won and used for the progress of all people. For space science, like nuclear science and all technology, has no conscience of its own. Whether it will become a force for good or ill depends on man. . . .
>
> We choose to go to the moon in this decade and do the other things, not because they are easy, but because they are hard; because that goal will serve to organize and measure the best of our energies and skills, because that challenge is one that we are willing to accept, one we are unwilling to postpone, and one which we intend to win.

It is hard to imagine a modern-day politician making a similar speech. Sometimes it seems that in the intervening years we have lost sight of how science progresses and grown intolerant of anyone who chooses to sail on a sea just because there might be new knowledge to be gained. We have developed an insatiable desire for short-term financial benefits and compelling newspaper headlines. We regularly vilify scientists whose work we deem, in our opinion as Google-qualified experts, to be impractical.

The human body is astoundingly complex, and no one really understands it all. The state of our knowledge is constantly evolving, and whatever views we hold on food and health, the reality is that eventually everyone will be proved wrong to some degree. Real scientists will always accept this, because they know that science, with its doubt and uncertainty, will at least be constantly moving closer to the truth. We need to remember that although certainty sells, only doubt can change the world. This leads us to **the final rule in the Angry Chef's Guide to Spotting Bullshit in the World of Food**: *They will tell you all these things with great certainty.*

Chapter Twenty-One

FIGHTING PSEUDOSCIENCE

*The fact that an opinion has been widely held is no evidence
whatever that it is not utterly absurd; indeed in view of the silliness of
the majority of mankind, a widespread belief is more likely
to be foolish than sensible.*

—BERTRAND RUSSELL

*A lie is halfway around the world before the truth
has a chance to get its pants on.*

—PERHAPS APPROPRIATELY THE SOURCE OF THIS QUOTE IS
DISPUTED, OFTEN ATTRIBUTED TO EITHER MARK TWAIN OR
WINSTON CHURCHILL. EITHER WAY IT IS A LITTLE OUTDATED,
AS THESE DAYS LIES CAN TRAVEL TWELVE TIMES AROUND THE
WORLD BEFORE THE TRUTH OPENS ITS EYES.

FIGHTING PSEUDOSCIENCE: AN APOLOGY

I think I might have made a terrible mistake. You see, when it
comes to fighting against pseudoscience and false beliefs, one of
the most important things to remember is not to draw attention
to the myths. In the excellent *Debunking Handbook*, John Cook
and Stephan Lewandowsky summarize a great deal of psychological research to produce a simple, practical guide to reducing the
impact of myths and false beliefs.[1] In it they outline a number of
effects, including the slightly surprising fact that debunking myths
can sometimes backfire by drawing them to people's attention.

Engaging with idiots can bring them into the debate, and when the people doing the engaging are academics or medical professionals, it can even legitimize the arguments. When writing about pseudoscience, you should certainly never headline the chapters of your book with specific myths . . .

Whoops.

. . . and instead just communicate the core facts that you know to be true. So, rather than saying "Vaccines don't cause autism," you should say, "Vaccines are safe and effective and prevent a number of terrible diseases." You should also try to refrain from too much detail and scientific terminology when debunking myths . . .

You mean go easy on the statistics.

. . . because simple falsehoods are often more appealing than a complex reality. In addition, you should take care to ensure the alternatives that you are proposing are not too difficult or impenetrable, or they might well be rejected.

Most of all, if you are going to wrench a long-held belief out of people's minds, then you need to have something to replace it that doesn't leave any gaps. People prefer an incorrect model to an incomplete one, so if your explanation does not explain all the observed reality, then it is likely to be rejected. If nonceliacs have crippling stomach pains, and these are relieved when they switch to a gluten-free diet, it is not enough just to say that they are imagining it. Their pains are real, and the relief is real, too, so telling them they are making stuff up, and perhaps even ridiculing them for doing so, is likely to be a highly counterproductive strategy. You need to provide a simple alternative explanation that does not patronize them, but also be careful not to provide a more complex solution. For instance, suggesting multiple FODMAP*

** Fermentable, **O**ligo-, **D**i-, **M**ono-saccharides, **A**nd **P**olyols. FODMAPs are chemicals that are naturally present in many foods, and it is believed that some people could be sensitive to them, causing a number of digestive problems.*

sensitivity—needing extensive investigation from a dietitian, followed by a structured plan over several months to investigate the actual cause of the problem—may just send them back to their gluten-free simplicity, which "just works."

Knowingly or not, pseudoscience movements tend to exploit gaps in the scientific consensus, so science really needs to provide a united front. Never is this more apparent than in the debate on obesity, where there is little agreement on the actual "cause" of this profound health crisis. This can lead to many peddlers of pseudoscience being believed when they mistake anecdote for evidence and claim that they have the answer. The moment cracks appear the weeds of pseudoscience grow between them, and if they are allowed to grow big enough there is a danger that they will strangle the voices of reason for ever. Clean eating, detox, the alkaline diet, Paleo, and many others have all grown between the cracks created by academics' arguing, disagreeing, and saying, "Well, it's complicated."

THE BIG PROBLEM

And therein lies the problem. Science needs to provide a consensus, yet the very engine that drives it is debate and disagreement. The moment it reaches complete accord, that is when progress ends. Demanding that the world of science present a united front is at odds with its very ethos. As we know from Science Columbo, it will always be bothered by one more thing.

Science also needs to provide clear, full, simple answers, and yet if it has "real" answers, then they are rarely complete and are

Restriction of FODMAPs has been found to be an effective treatment for irritable bowel syndrome for some people. FODMAP restriction should only be carried out under the supervision of a dietitian, as a number of foods must be excluded initially and then slowly reintroduced to find the exact nature of the sensitivity. The initial exclusion diet is likely to be highly restrictive and if carried out without professional supervision could potentially be extremely harmful.

generally riddled with uncertainty and full of doubt. The more that nutrition is studied, the more there is to know, with most areas revealing greater complexity as they are investigated. Without being disingenuous and misleading people, science cannot provide the compelling answers and clearly defined rules that our strangely wired brains desire. It can provide opinions and offer suggestions, but it will never say that it definitely knows.

Pseudoscience has a huge advantage because it can appeal to our instinctive brains and create easy-to-follow rules without the handicap of having to be right. It will happily classify things as black and white, good and bad, clean or contaminated. Science cannot, and should not, do any of these things. And so it is trapped by the truth, always destined to spin an inferior-sounding story.

To make matters worse, as we previously discussed, attitudes to food can be full of "protected values," things that are believed absolutely and are insensitive to evidence, meaning that no amount of reasoning and evidence will sway these people's beliefs.[2] Commonly recommended persuasion techniques, such as demonstrating clear arguments or recommending evidence-based alternatives, will simply not work in these cases.

THE CHANGING WORLD OF INFORMATION

When we "do some research," just as our gluten-free experimenter Jamie did at the beginning of the book, we discover knowledge for ourselves. This search is riven with the potential for bias and inaccuracies, especially as our search engines and social media feeds become more tailored to show us only the sort of material we desire. Millions of pieces of information will be listed in a specific order, and when an item is at the top of this list we will be inclined to believe that it is the most important, the most relevant, and the most truthful. Increasingly, this order has more to do with our previous search histories, likes, and interests than the quality of the evidence. We feel that we have ownership of any answers we have discovered for ourselves, giving them infinitely more power than anything delivered by the NIH or a CDC press release.

Pseudoscience, conspiracy theories, and the like all flourish in the modern information age because it is incredibly easy for extreme messages to reach the public and become shared. Information is now shared horizontally, rather than disseminated by trusted authorities. The way we interact with and seek information is constantly changing, and the potential for the existence of pockets of strange beliefs increases every day. The internet is big enough to harbor closed social media communities full of like-minded people, and dangerous false beliefs have the ability to flourish and grow completely unchecked, giving them the potential to do enormous harm.

THE PROBLEM WITH MEDICINE

Britt Marie Hermes, the converted naturopath we discussed earlier, spent many years treating patients who came to her with real problems affecting their lives. She says:

> The naturopathic business model is far more simple than that of a conventional medical practice. A lot of the training covers how to develop patient relationships and offer what seems like personalized care. Naturopaths can't afford to hire staff, but this means that patients are not just shoved around from room to room by nurses. The naturopath does everything from greeting the patient at the front desk, taking vitals and scheduling follow-up appointments. You develop a really good relationship with them.
>
> Lots of people came in with self-diagnosed conditions like multiple chemical insensitivity, chronic fatigue or

chronic Lyme when it was clear the underlying cause might be depression. People really like the personalized feel of the naturopathic practice and the "treatments" often involved a lot of counselling, which genuinely made them feel like they were being heard. A lot of patients are very sensitive about being told they are depressed and view conventional medicine as dismissive. Because there is a stigma about mental health problems, patients focus on the physical symptoms, such as fatigue. Such patients are attracted towards naturopathic treatments.

Some of the blame for this has to go to the traditional medical establishment, with its tendency to "diagnose and dump" that we saw previously, and with its dismissive attitude toward certain conditions. Conventional medicine does not deal well with ambiguity, when there is no clear pathway from GP to specialist to treatment. But we need to maintain realism and understand that properly qualified health professionals will always be in huge demand, never able to compete with the sort of integrative service that naturopaths or "nutritional therapists" can offer. When dietitians say that only they should be trusted when it comes to offering advice on diet and health, they do need to appreciate that there are not enough dietitians to go around, and many people are likely to end up following a different path.

To fight effectively, we also need to look at our society's attitude toward mental illness. We must put an end to the stigmatization of depression, know what a debilitating condition it can be, and realize that the consequences of dismissing it can be devastating. Troubled attitudes to food and health are often symptoms of this condition, and the more we can understand that relationship, the more we can help. Many of the symptoms that health bloggers discuss in relation to detox diets, alkaline, clean eating, Paleo, or even GAPS are common symptoms of depression, and for people who are vulnerable, desperate for help, and unable to accept their underlying condition, there is a danger of being driven to any port in a storm. Only by breaking the stigma can we prevent this from happening, bringing sufferers in from the darkness and providing genuine, effective treatments. If we don't, we risk pushing them into the arms of pseudoscience, creating troubled relationships with food, restricted diets, illness, and suffering that will compound the problems they already face.

THE CURSE OF KNOWLEDGE

To make matters even worse, many scientists are afflicted with the curse of knowledge. Anyone who has ever played the game where you tap out the rhythm of a piece of music and ask someone to guess the tune will be aware of this problem. In a much-cited experiment conducted in 1990, Stanford graduate student Elizabeth Newton showed that the guesser only gets the correct tune 2.5 percent of the time, but the tapper believes that they will guess correctly on 50 percent of occasions. When you have the tune playing happily in your head it seems obvious and intuitive what the answer is, but when all you can hear is a series of taps, it is nearly impossible. This is the curse of knowledge. Scientists seem incredulous when their arguments are doubted in favor of some new-age health guru, internet-qualified nutritionist, or celebrity bandwagon-jumper. To the scientist, the need for experimentation, trial, peer review, high-quality evidence, and representative samples is obvious; it is the tune that plays constantly in their head. But they need to understand that it is a tune that only they can hear. Not only do most of us not understand the science itself, we also fail to understand the method, the reasons why science holds the key. It is sometimes tough for scientists to appreciate that normal people prefer anecdotes and simple rules and are willing to believe them even though they are wrong.

WHAT TO DO?

Science is unsuited to providing sensible, coherent messages and is defined as much by what it doesn't know as what it does. The medical establishment, with its massive pressures and demands, unknowingly pushes away many people in need, sending them into the grateful arms of many a charlatan. The way we access information is tailor-made for false beliefs to grow and persist. The more society becomes educated, the more individualistic people become, ironically propelling them toward ideas that challenge the mainstream. We are doomed, and the false prophets will surely win, forcing us into relationships with food that are brutal, restrictive, and damaging—and will harm us all.

Thank you for reading this book. The End.

Or maybe not. I can do nothing about the way science is communicated, and I understand that for serious scientists to engage with the sort of dietary woo that I have discussed in this book would be to legitimize it and draw it into the debate. But this means that the rats of pseudoscience are often allowed to run amok, and in this world of fragmented information, they can do great harm.

What I can do is poke fun at them. I can ridicule their stupidity. For I am no expert; I am just a chef. This is exactly the reason the Angry Chef was born. He is a character, created for the sole purpose of fighting back. I may write his words, but he is not real. He is the product of a number of helpers, scientists, dietitians, reference checkers, and interested others who contribute knowledge and information to the cause. He is just one small weapon designed to fight pseudoscience in food, and he neglects to use any of the considered and careful techniques for debunking from Cook and Lewandowsky's handbook—for a number of important reasons.

I believe that the key to believing any message is our relationship with those who deliver it. Sometimes scientists, cursed by their knowledge and tainted by a history of scientific transgressions, can seem distant, arrogant, and out of touch with reality. Angry Chef is just one alternative voice that might appeal to a few people. His expletive-laden rants, his twisted and often peculiarly British sense of humor (apologies for any of that which is lost in translation), his unfashionable respect for scientific consensus and public health messages, his distaste for cognitive ease and self-serving celebrities; these things appeal to some people, cutting through the bullshit and providing a tiny oasis of sense. I get a lot of correspondence, much of it not very polite, but occasionally someone will contact me with a touching and thought-provoking comment. Below are some examples:

Every time I take a bite of pizza or nibble on a cookie or consume anything from the dreaded dairy family, I feel the horrendous guilt and shame come upon me and it's a daily fight to love my food. . . . But then I stumbled upon your rants—your fantastic, humour-filled, rage-filled, cuss-filled rants—and I took a very deep breath of air because you are saying things that I have so desperately wanted to say for a very long time, have felt very deep inside, but have been so fearful to embrace. To do so would mean to exclude myself from a world that I have been part of for so long. It would also mean surrendering myself to the enjoyment of food, rather than the obsessive need to prioritize health over flavour. I just wanted to say thank you, because it's been a long time since I could breathe.

Here's another contributor, in response to a post on clean eating:

We've gone seriously wrong somewhere along the line to have ended up with a society where a problem of this hor-ror, scale and severity both exists and is largely ignored. And it's getting worse. And will continue to get worse unless something changes. So thank you for putting your name and platform to it.

And this in response to a piece on eating disorders:

I've recently been tested for food intolerance and to my joy have been given a long list of "do not eats." Hooray, at last, an excuse to refuse myself. Your timely post has come to me at a crossroads and common sense has slapped me in the face and for that I thank you.

This reaction is not bad for an imaginary chef who talks to a voice in his head. Angry Chef might only appeal to a few people, his voice(s) might only reach one or two, but the reason he takes up so much of my life is because he makes a tiny difference in some people's lives. He takes away some of the confusion and exposes some of the lies.

When the gurus of nutri-nonsense have to argue with an imaginary chef, they are revealed for what they are: ridiculous, bizarre, foolish, and wrong. But Angry Chef will not appeal to everyone. He needs your help to fight the same battle in different ways. If a thousand people can be persuaded to shout the same message in different voices, then there is even more of a chance that we might hold back the tide of dietary bullshit. We need strong, passionate advocates in every type of media, appealing to every age group and demographic. Social proof drives the success of fad diets, and so social proof has the potential to destroy them.

HOW SCIENCE SHOULD BE TAUGHT

How might this happen? Maybe when science is taught in school, it should not be presented as a list of facts. You do not teach someone art by providing information about paint. Science is about profound and interesting things, about investigation, search, truth, and evidence. Science should teach children to doubt, to question, and to understand the wisdom of knowing your own ignorance. Although facts are important, the output of school science teaching should not be children who know about photosynthesis, thermodynamics, and semipermeable membranes. We should be trying to produce children who understand that correlation is not always causation, that anecdotes are not evidence, that a theory is not something dreamed up in the bar, and that interesting results are often wrong.

Science education should teach us what regression to the mean is, but also how easily it can fool us. It should explain how inclined we are to find patterns in randomness, how drawn we are to accept rules and certainty, and how our instinctive brain guides many decisions we make. It should produce adults who are smart enough to see that their perspective is limited, and that they can be fooled. We should be taught to think, but also be taught how we think. Science should spend more time explaining how our brain can trick us into false beliefs and then reveal how the scientific method saw through this and changed the world.

The facts produced by science will always change, sometimes fundamentally, but what will remain the same is the way that it works and the principles that make it the greatest force for progress that there has ever been. If taught well, these could stay with us for life and make us stronger for it.

If we can get this right, then there will be voices. There will be smart, eloquent, informed people who know how to spot the charlatans when they appear. They will shout them down, they will convince their friends, they will be so vociferous and determined that moments of doubt will be created for everyone. If an understanding of evidence is built into every science education, if the desire to question and doubt is instilled in schools, if it can be taught that understanding the gaps in our knowledge is the true measure of our intelligence, then perhaps the bullshit will not take hold. Together, we will shine bright lights on the rats of pseudoscience and watch them scatter away into the dark.

Epilogue

Fools and fanatics are always so certain of themselves,
but wiser men so full of doubts.

—BERTRAND RUSSELL

T hank you for reading this far.

I mentioned right at the beginning that you would not find the hidden secret to healthy eating within the pages of this book and I have tried not to give out too much advice on diet, lest I should fall into the trap of preaching beliefs based on my own cognitive bias. But I also said at the beginning that we cannot choose the food we eat with impunity, as there are clearly links between diet and health. So, in this final section, I would like to tell you what I know about how to eat healthily. Don't worry; it really won't take long.

Although I have written a book without a single recipe, I have been a professional chef for over twenty years, and that career has been driven by a profound love of food and cooking. And although I love to share this passion, I believe that the world has enough cookbooks already, so it has never been my desire to add one more. I am also slightly afraid that if I did ever write a book of recipes, the publishers might ask me to include a section explaining my food philosophy, at which point I would have to punch them in the face.

My desire to write this book was, however, still driven by my passion for food. Our relationship with the food we eat is extremely important, in many ways defining what it means to be human, and I have always had a burning desire for people to have as much of a joyous relationship with food as I do. If you can derive great joy from something you have to do several times a day, then it will add a lot of richness and fulfillment to your life. I work as a chef because I want to help people enjoy the food they eat, and so, in some small way, improve the quality of their lives. This book was driven by the same desire, to cut away some of the bullshit that pervades the world of food and enable people to enjoy each mouthful a little more.

Eating well is about pleasure, balance, and the creation of memories. It is a fine way to embellish the most important moments of our lives, adding richness and texture to precious times and enhancing moments of joy. The more we break food down, the more we try to define it by the chemicals it contains, the more we label items as clean, good, immoral, contaminated, or pure, and the more we attach guilt or shame to our choices, the further away we get from the sensible, balanced relationship that we need.

We need to recognize and understand that when it comes to the food we eat, control and restriction are not enviable moral qualities. A positive relationship with food comes from embracing variety and achieving balance. The joy we can find in food and cookery is more powerful than any antioxidant or phytochemical when it comes to improving our health and well-being. Poor dietary choices do not occur when people are driven by hedonistic pleasure; they occur when people eat without thought, and that will never happen if we engage with and truly love the food we eat.

Captain Science, my anonymous collaborator and provider of references and research, once said something very wise to me regarding nutrition. She has dedicated years of her life to studying dietetics and biological sciences, including time carrying out many of the sort of reductionist experiments that dominate the agenda, and has eventually come to the conclusion that "the more you learn about diet, the less interesting the final message is: Eat everything in moderation and move around more."

When it comes to understanding how food affects our health, science has made astounding progress, but there are still many gaps in our knowledge. Usually, the most that science can truly tell us is that the tens of thousands of chemical substances that we ingest every day as food act together in miraculous and poorly understood combination. It is our job to live with this uncertainty and try to make reasonable judgments anyway. Try to be immune to the charms of easy narratives and false hopes, because at best they are wrongheaded, and at worst, truly dangerous. I do hope that this book has left you full of uncertainty and doubt and a little wiser for it.

So, given everything I have learned over the years, I would like to present . . .

THE ANGRY CHEF'S VERY BRIEF GUIDE TO EATING WELL

Since the real science behind food is still driven by a lot of uncertainty as to how our diet affects our health, the best that we can do is to embrace as much variety as possible. Unless your doctor has recommended steering clear of something specific, don't restrict anything from your diet, and try new things all the time. If you can learn to love and engage more with the food you eat, this will be easy to do. Try to learn a few cooking skills and spend a bit of time in the kitchen if you can. If nothing else, once you have learned enough to do it instinctively, it provides a good bit of thinking time every day.

But if you don't enjoy cooking, and I am aware that many people do not, there is no need to feel guilty about it. Eating takeout, a package of cookies, chicken nuggets, or a TV dinner now and again will not harm you. Neither will a bowl of kale and quinoa salad. Just make sure that this is not all you ever eat. The only diets that can harm your health are diets of restricted choice, lacking in variety and devoid of joy.

So, the secret to healthy eating that I mentioned back in chapter 10 is here:

1. Eat lots of different stuff.

2. Not too much or too little.

3. Try to achieve a bit of balance.

4. Try not to feel guilty. Most important, never make others feel guilt or shame about the food they eat.

That's it. I apologize for taking so long to get around to telling you. It's not the path to a miracle cure for disease or eternal life, or even for shiny hair, but it is the best advice we have.

Oh, there's one more thing. Try to eat a little fish now and again, especially oily fish, such as salmon or mackerel. With all the educated and informed people I have spoken to, all the scientific opinions I have gathered, and everything I have read, about the only thing that everyone can agree on is that incorporating a bit of fish into your diet could be of some benefit. Maybe. Well, it isn't going to do much harm if you do, and perhaps more importantly, if you learn to cook it properly it is one of the most delicious foods on earth.

Maybe we should do a fish cookbook.

Shut up.

Thank you for reading. Eat well.

Appendix 1

THE ANGRY CHEF'S GUIDE TO SPOTTING BULLSHIT IN THE WORLD OF FOOD

1. They will have a food philosophy.

2. They will try to sell you detox.

3. They will tell you that your illness is your own fault.

4. They will fit the health-blogger template.*

5. They will tell you that the truth is not up for debate.

6. They will talk about superfoods.

7. They will use anecdotes as evidence.

8. They will quote ancient wisdoms at you as fact.

9. They will tell you things were better "back then."

10. They will tell you all these things with great certainty.

* "I was living my impossibly glamorous life as an **INSERT GLAMOROUS OCCUPATION HERE** at a hundred miles an hour, eating all sorts of junk and not caring what I put in my body. My health was really suffering. It was only when I started to take control of the food I was eating that my health improved. I started my **INSERT NAME OF MADE-UP DIET PLAN HERE** and it revolutionized my life. All my friends just begged me to share my recipes with them, and that's how my blog was born."

Appendix 2

THE ANGRY CHEF'S SIMPLE GUIDE TO WHO WE SHOULD BELIEVE IN THE WORLD OF FOOD

Registered Dietitian (RD)—This is always a good place to start. The term *dietitian* is protected by law, so if someone uses this term to describe themselves, it means they are qualified to use the science of nutrition to devise eating plans for patients. A dietitian, also signified by the initials RD after their name, will have studied for a degree in dietetics and are registered healthcare professionals. Qualification to be a dietitian is earned either through a four-year degree program or through a graduate program combined with a supervised twelve hundred-hour internship in a hospital or private practice. And as nutrition is a fast-moving science, dietitians are also required to regularly update and maintain their credentials in order to keep their registration. In my experience they are almost all dedicated, professional, and intelligent and give out the best evidence-based information available. The only negative is that there are just not enough of them to go round.

Certified Nutrition Specialist (CNS)—Although many medical doctors spend a lot of time talking about diet and health, they are not always to be trusted as they learn very little about nutrition in medical school. A few take an extra certification to become a Certified Nutrition Specialist, and can use the initials CNS after their name. This means they can do many of the same things dietitians do, and are probably well qualified to give out diet and health advice. CNS doctors are a rare breed, however, and most MDs who comment about nutrition in the media don't seem to have this qualification. They obviously prefer their instincts and intuition over evidence-based advice.

Nutritionist—This is where the problems start. Whereas dietitian is a legally protected term, nutritionist is not. Technically anyone can call themselves a nutritionist, and many people do. If you are reading this, then by being a human who can say words, you already have all the qualifications required to call yourself a nutritionist. Really. Get a fancy sign printed up for your door, stick an ad in the local newsstand, and start taking consultations. Give it a few weeks and media outlets will probably start asking you for your hot takes on nutritional topics.

Certified Nutritionist (CN)—It is a different matter if you want to call yourself a certified nutritionist, as many of the current breed of internet-based health gurus do. You will need to obtain a nutritionist certificate. Fortunately, this need not be much of a hindrance to your newfound career, as a nutritional qualification can be obtained with a few hours of online study for about $20. Congratulations, you may now get a new sign made and put up your prices.

Nutritional therapist—A title favored by many health bloggers and clean eaters, perhaps driven by an understanding that the sort of authority conveyed by sciencey-sounding terms like nutritionist is a bit outdated. "Nutritional therapist" is distinctly more new age and holistic sounding. Of course, it is just a made-up thing, and the term is not protected in any way. We are all nutritional therapists. There are a couple of bodies that give accreditation to some courses, mostly distance learning, short-course type things. One of these is the General Regulatory Council for Complementary Therapies, and it is well worth keeping in mind that these people also cover therapies such as crystal healing. Here, where I am, the British Association for Applied Nutrition and Nutritional Therapy (BANT) is a professional association of which many nutritional therapists are members. Let's just say that a number of GAPS diet practitioners are also registered with BANT, which should tell you everything you need to know about their requirement for evidence-based practice.

Nutrition experts—It does amuse me greatly when I see someone making comments in the media about food and health matters when they describe themselves as a "nutrition expert." I would suggest that anyone who feels the need to call themselves an expert in anything probably isn't one. Other titles to look out for are healthy-eating expert, naturopathic nutritionist, naturopath, holistic food healing expert, well-being expert, and mindful nutrition specialist. By the time I have written this book, there will probably be dozens more spurious titles created with blogs, Instagram accounts, and bowls full of kale, all clamoring for the attention of newspapers and magazines. Hopefully now you will be better equipped to sniff them out when they do.

I am sure there are some decent unregistered nutritionists, nutritional therapists, and self-styled nutrition experts out there, giving out sensible, evidence-based information and helping people to eat balanced diets, but unfortunately I have not come across one yet. In a busy world, where shortcuts are needed to help our instinctive brain make useful judgments, I would suggest focusing on whether or not you see the initials RD or CNS after someone's name, and, if not, take their advice with a pinch of salt. Just make sure that pinch contains less than 20 percent of your recommended daily allowance (RDA) of sodium.

There are a number of other titles conveying authority on matters of diet and nutrition. There are professors and doctors of nutritional and life sciences; there are medical doctors (who, contrary to the claims of many naturopathic practitioners, do have a knowledge of nutrition and are qualified to give advice); and there are many others in the world of respectable academia with vast knowledge of food science and health. Unfortunately, as we have discussed, when it comes to the more specialized academics, and even some people within the medical profession, we cannot always trust what we hear.

OK. So, I look for RD or CNS and ignore everything else. I can do that.

Good. That will make things much easier.

Why are you so sure dietitians are OK? I heard they were all paid shills for the food manufacturing industry.

Because they base their advice on evidence-based scientific research and the sort of systematic reviews of evidence that get us as close to the truth as possible. It is not perfect; sometimes it is proven to be wrong, but it is as good as we are likely to get. I hope throughout this book we have at least helped explain that although the messages of science are not always the most instinctively appealing, there are many good reasons why we should believe them.

Acknowledgments

I have always considered *The Angry Chef* as far from being my work alone. I am just a chef with an interest in various aspects of food. Without the knowledge, insight, and experiences of many others, this book would be nothing but the ramblings of a grumpy, middle-aged man. There are many people whom I need to thank for contributing, supporting, and helping over the last year or so.

First, Captain Science, my anonymous and brilliant collaborator who keeps me on the straight and narrow and is perhaps the greatest checker of facts that I know. When Captain Science appeared in the world of Angry Chef, it changed from being a succession of incoherent and demented rants into something far more interesting. With references.

Incredibly though, Captain Science does not know everything, and many others helped hugely to cover any gaps in her knowledge. Here is a list of just a few, with apologies to those I might have missed. Judy Swift for her help, insight, and passion (for legal reasons I can only report about 10 percent of what she tells me—I wish I could publish more). Peter Herman, Steven Pinker, and Paul Rozin for being so generous with their valuable time and knowledge. Helen Lynn, Luise Kloster, and Richard Johnson for their early and essential backing. Alan Marson and Chris Marson for everything that they did to help. Simon Godsell for a drawing that brought Angry Chef to life in my mind. Helen West, Jennifer Low, and Catherine Collins for their time, information, references, and constant support. Emily-Rose Eastrop from IFHP for her insights and challenges. Ian Marber and Britt Marie Hermes for their support, time, and unique inside perspectives. David Speigelhalter for his crusading knowledge and passion. Sydney Scott for her help and studies. David Gorski for his inspiringly long and detailed blog posts. David Robert Grimes for his knowledge, passion, and help on the subjects of both cancer and physics. Alice Roberts and Mark Thomas for their insights into the past (and Alice for supporting Angry Chef from the very beginning). Chris Peters from Sense About Science for his support and information. Michele Belot and Jeff Brunstrom for insight into behavior and nudges. Renee McGregor for being a constant and valuable resource in so many areas. Zoe Connor for her inspiring anger, passion, and dedication. Jane Smith from ABC for her incredible, crusading work. Eric Johnson-Sabine for his help. Everyone at Cancer Research and Research Autism for information and help. Annie Grey for her remarkable knowledge of the past and constant support of my writing. Rachel Laudan for her time and brilliance. Javier Gonzalez and Chris Starmer for their help and stories. Claire Marriot for a huge amount of invaluable support and for being a great sounding board for ideas. Aaron Calvert for his time and unusual perspective. Melisa (Autistic Zebra) for being a huge inspiration and long-term supporter. Anna and Emma for all their help, bravery, and honesty. Emily Sterling for her contacts and interest. Eve Simmons and Laura Dennison from Not Plant Based for everything they have done. Jon White, Lucy Dunn, Susan Low, and Amanda Ursell, without whom I would probably still just be talking to myself. Alexandra Cliff, without whom I would definitely just be talking to myself, and for making me believe that I might be able to write a bit. Alex Christofi, who made this book real, created the bullshit octopus, and added a huge amount—if ever you need some unpaid work editing blog posts, you know where I am. And to Nottingham Forest, who were so dismally poor at the beginning of the 2016/17 season that I had no problem dedicating so much time to my writing.

Many others have supported and contributed to the Angry Chef website and blog. It is a constant surprise and joy that so many people find the time to read my words and are inspired to interact. I can only thank you all.

Lastly, a few special ones. Ellie for inspiring all that I do. My family for their time, interest, and grudging acceptance of my profanity. And most of all, to Mrs. Angry Chef, for an endless supply of tea and understanding. Every word was for you.

Notes

PART I: GATEWAY PSEUDOSCIENCE
CHAPTER 2: DETOX DIETS
1. "Joint FAO/WHO Expert Committee on Food Additives, 72nd Meeting: Summary and Conclusions," fao.org/3/a-at868e.pdf.
2. "Joint FAO/WHO Expert Committee on Food Additives 74th Meeting: Summary and Conclusions," fao.org/3/a-i2358e.pdf.
3. Jérôme Ruzzin, "Public Health Concern Behind the Exposure to Persistent Organic Pollutants and the Risk of Metabolic Diseases," *BMC Public Health* (April 20, 2012), ncbi.nlm.nih.gov/pmc/articles/PMC3408385; see also UNEP "United Nations Environment Programme: State of the Science of Endocrine Disrupting Chemicals" (2012), who.lnt/ceh/publications/endocrine/en.
4. H. Ren et al., "Effect of Chinese Parsley *Coriandrum sativum* and Chitosan on Inhibiting the Accumulation of Cadmium in Cultured Rainbow Trout *Oncorhynchus mykiss*," *Fisheries Science* 72, no. 2 (2006): 263–69, link .springer.com/article/10.1111/j.1444- 2906.2006.01147.x; see also M. Aga et al., "Preventive Effect of Coriandrum Sativum (Chinese Parsley) on Localized Lead Deposition in ICR Mice," *Journal of Ethnopharmacology* 77, no. 2–3 (2001): 203–8, sciencedirect.com/science/article/pii/S0378874101002999.
5. J. Allen et al., "Detoxification in Naturopathic Medicine: A Survey," *Journal of Alternative and Complementary Medicine* 17, no. 12 (December 2011): 1175–80, ncbi.nlm.nih.gov/pmc/articles/PMC3239317.
6. General—A. V. Klein and H. Kiat, "Detox Diets for Toxin Elimination and Weight Management: A Critical Review of the Evidence," *Journal of Human Nutrition and Dietetics* (December 18, 2012), onlinelibrary.wiley.com/ doi/10.1111/jhn.12286/abstract.

CHAPTER 3: THE ALKALINE DIET
1. H. C. Sherman and A. O. Gettler, "The Balance of Acid-Forming and Base-Forming Elements in Foods and Its Relation to Ammonia Metabolism," Journal of Biological Chemistry 11 (1912): 323–38, jbc.org/content/11/4/323.full .pdf?sid=a21d1cda-b55d-4218-bea6-0a9f023dbfbe.
2. T. Buclin et al., "Diet Acids and Alkalis Influence Calcium Retention in Bone," *Osteoporosis International* 12, no. 6 (2001): 493–99, ncbi.nlm.nih.gov/ pubmed/11446566.
3. T. Remer and F. Manz, "Estimation of the Renal Net Acid Excretion by Adults Consuming Diets Containing Variable Amounts of Protein," *American Journal of Clinical Nutrition* 59 (1994): 1356–61,nutritionj.biomedcentral.com/ articles/10.1186/1475-2891-8-41.
4. R. P. Heaney and K. Rafferty, "Carbonated Beverages and Urinary Calcium Excretion," *American Journal of Clinical Nutrition* 74 (2001): 343–47, ajcn .nutrition.org/content/74/3/343.long.
5. T. R. Fenton et al., "Phosphate Decreases Urine Calcium and Increases Calcium Balance: A Meta-Analysis of the Osteoporosis Acid–Ash Diet Hypothesis," *Nutrition Journal* 8, no. 41 (2009), nutritionj.biomedcentral.com/ articles/10.1186/1475-2891-8-41.

PART II: WHEN SCIENCE GOES WRONG
CHAPTER 7: COCONUT OIL

1. L. Hooper et al., "Effect of Cutting Down on the Saturated Fat We Eat on Our Risk of Heart Disease," *Cochrane Review* (2015), cochrane.org/CD011737/VASC_effect-of-cutting-down-on-the-saturated-fat-we-eat-on-our-risk-of-heart-disease.

2. Russell J. de Souza and Andrew Mente, "Intake of Saturated and Trans Unsaturated Fatty Acids and Risk of All Cause Mortality, Cardiovascular Disease, and Type 2 Diabetes: Systematic Review and Meta-Analysis of Observational Studies," *British Medical Journal* (2015): 351:h3978, bmj.com/content/351/bmj.h3978.long.

3. T. S. T. Mansor et al., "Physicochemical Properties of Virgin Coconut Oil Extracted from Different Processing Methods," *International Food Research Journal* 19, no. 3 (2012): 837–45, ifrj.upm.edu.my/19 (03) 2012/(8) IFRJ 19 (03) 2012 Che Man.pdf.

4. Jon J. Kabara et al., "Fatty Acids and Derivatives as Antimicrobial Agents," *Antimicrobial Agents and Chemotherapy* 2, no. 1 (1972): 23–28, ncbi.nlm.nih.gov/pmc/articles/PMC444260.

5. Poonam Sood et al., "Comparative Efficacy of Oil Pulling and Chlorhexidine on Oral Malodor: A Randomized Controlled Trial," *Journal of Clinical and Diagnostic Research* 8, no. 11 (2014), ncbi.nlm.nih.gov/pmc/articles/PMC4290321.

6. A. G. Dulloo et al., "Twenty-Four-Hour Energy Expenditure and Urinary Catecholamines of Humans Consuming Low-to-Moderate Amounts of Medium-Chain Triglycerides: A Dose-Response Study in a Human Respiratory Chamber," *European Journal of Clinical Nutrition* 50, no. 3 (1996): 152–58, ncbi.nlm.nih.gov/pubmed/8654328.

7. A. Prior et al., "Cholesterol, Coconuts, and Diet on Polynesian Atolls: A Natural Experiment: The Pukapuka and Tokelau Island Studies," *American Journal of Clinical Nutrition* 34, no. 8 (1981): 1552–61, ncbi.nlm.nih.gov/pubmed/7270479.

8. S. Mendis, U. Samarajeewa, and R. O. Thattil, "Coconut Fat and Serum Lipoproteins: Effects of Partial Replacement with Unsaturated Fats," *British Journal of Nutrition* 85, no. 5 (2001): 583–89, cambridge.org/core/journals/british-journal-of-nutrition/article/coconut-fat-and-serum-lipoproteins-effects-of-partial-replacement-with-unsaturated-fats/1793E90524FC34B3626 1581F36676E72.

9. World Health Organization GHO data, who.int/gho/ncd/risk_factors/cholesterol_text/en; see also: Prospective Studies Collaboration, S. Lewington et al., "Blood Cholesterol and Vascular Mortality by Age, Sex, and Blood Pressure: A Meta-Analysis of Individual Data from 61 Prospective Studies with 55,000 Vascular Deaths," *Lancet* 370, no. 9602 (December 1, 2007): 1829–39, ncbi.nlm.nih.gov/pubmed/18061058.

10. Kai Ming Liau et al., "An Open-Label Pilot Study to Assess the Efficacy and Safety of Virgin Coconut Oil in Reducing Visceral Adiposity," *ISRN Pharmacology* (2011), ncbi.nlm.nih.gov/pmc/articles/PMC3226242.

11. General—Dr. Laurence Eyres, "Coconut and the Heart Evidence Paper," NZ Heart Foundation, 2014.

CHAPTER 8: THE PALEO DIET

1. P. Gerbault et al., "Evolution of Lactase Persistence: An Example of Human Niche Construction," *Philosophical Transactions of the Royal Society B: Biological Sciences* 366, no. 1566 (March 27, 2011): 863–77, ncbi.nlm.nih.gov/pmc/articles/PMC3048992.

2. Karen Hardy et al., "The Importance of Dietary Carbohydrate in Human Evolution," *Quarterly Review of Biology* 90, no. 3 (September 2015): 251–68, jstor.org/stable/10.1086/682587?seq=1 page_scan_tab_contents.

3. L. Hooper et al., "Effect of Cutting Down on the Saturated Fat We Eat on Our Risk of Heart Disease," *Cochrane Review* (2015), cochrane.org/CD011737/VASC_effect-of-cutting-down-on- the-saturated-fat-we-eat-on-our-risk-of-heart-disease.
4. Deirdre K. Tobias et al., "Effect of Low-Fat Diet Interventions Versus Other Diet Interventions on Long-Term Weight Change in Adults: A Systematic Review and Meta-Analysis," *Lancet* 3, no. 12 (2015): 968–79, thelancet.com/journals/landia/article/PIIS2213-8587(15)00367-8/fulltext.
5. Dr. David Nunan and Dr. Kamal R. Mahtani, Centre for Evidence-Based Medicine—CEBM response: "Eat Fat, Cut the Carbs and Avoid Snacking to Reverse Obesity and Type 2 Diabetes," cebm.net/centre-evidence-based-medicine-response-report-eat-fat-cut-carbs-avoid-snacking-reverse-obesity-type-2-diabetes.
6. M. Gleeson, "Immunological Aspects of Sport Nutrition," *Immunology and Cell Biology* 94, no. 2 (February 2016): 117–23, ncbi.nlm.nih.gov/m/pubmed/26634839/?i=2&from=carbohydrate immunity athletes.

CHAPTER 9: ANTIOXIDANTS

1. G. Bjelakovic et al., "Antioxidant Supplements for Prevention of Mortality in Healthy Participants and Patients with Various Diseases," *Cochrane Review*, (2012), ncbi.nlm.nih.gov/pubmed/22419320.
2. G. S. Omenn et al., "Effects of a Combination of Beta Carotene and Vitamin A on Lung Cancer and Cardiovascular Disease," *New England Journal of Medicine* 334, no. 18 (1996): 1150–55.
3. The Alpha-Tocopherol Beta Carotene Cancer Prevention Study Group, "The Effect of Vitamin E and Beta Carotene on the Incidence of Lung Cancer and Other Cancers in Male Smokers," *New England Journal of Medicine* 330 (1994): 1029–35.
4. Z. Schafer et al., "Antioxidant and Oncogene Rescue of Metabolic Defects Caused by Loss of Matrix Attachment," *Nature* 461, no. 7260 (2009): 109–13.
5. P. M. Kris-Etherton and C. L. Keen, "Evidence That the Antioxidant Flavonoids in Tea and Cocoa Are Beneficial for Cardiovascular Health," *Current Opinion in Lipidology* 13, no. 1 (2002): 41–49.

CHAPTER 10: SUGAR

1. R. B. Ervin and C. L. Ogden, *Consumption of Added Sugars Among U.S. Adults, 2005–2010*, NCHS data brief no. 122 (Hyattsville, MD: National Center for Health Statistics, 2013), cdc.gov/nchs/data/databriefs/db122 .htm#x2013;2010%20; ncbi.nlm.nih.gov/pubmed/22617043.
2. Scientific Advisory Committee on Nutrition, *Carbohydrates and Health* (London: TSO, 2015), www.gov.uk/government/publications/sacn-carbohydrates-and-health-report.
3. Alan W. Barclay and Jennie Brand-Miller, "The Australian Paradox: A Substantial Decline in Sugars Intake over the Same Timeframe that Overweight and Obesity Have Increased," *Nutrients* 3, no. 4 (2011): 491–504, mdpi.com/2072-6643/3/4/491/htm.
4. DEFRA, National Food Survey Data, http://webarchive. nationalarchives. gov.uk/20130103014432, defra.gov.uk/statistics/foodfarm/food/familyfood/nationalfoodsurvey.
5. B. A. Bates et al., eds., "National Diet and Nutrition Survey Results from Years 1, 2, 3 and 4 (Combined) of the Rolling Programme (2008/2009–2011/2012): A Survey Carried Out on Behalf of Public Health England and the Food Standards Agency," London, 2014.
6. British Heart Foundation, "Coronary Heart Disease Statistics," 2012, bhf.org.uk/publications/statistics/coronary-heart-disease-statistics-2012.

7. James M. Rippe and Theodore J. Angelopoulos, "Sugars and Health Controversies: What Does the Science Say?" *Advances in Nutrition* 6, no. 493S–503S (2015), advances.nutrition.org/content/6/4/493S.abstract.

8. C. S. Srinivasan, "Can Adherence to Dietary Guidelines Address Excess Caloric Intake? An Empirical Assessment for the UK," *Economics and Human Biology* 11, no. 4 (2013): 574–91, sciencedirect.com/science/article/pii/S1570677X13000385.

9. General—C. Snowdon, "The Fat Lie," Institute of Economic Affairs Report, 2014, iea.org.uk/publications/research/the-fat-lie.

PART III: THE INFLUENCE OF PSEUDOSCIENCE
CHAPTER 11: A HISTORY OF QUACKS

1. B. Bennett, "Doctrine of Signatures: An Explanation of Medicinal Plant Discovery or Dissemination of Knowledge?" *Economic Botany* 61, no, 3 (2007): 246–55.

CHAPTER 13: PROCESSED FOODS

1. Sydney E. Scott, Yoel Inbar, and Paul Rozin, "Evidence for Absolute Moral Opposition to Genetically Modified Food in the United States," *Perspectives on Psychological Science* 11, no. 3 (2016): 315–24, pps.sagepub.com/content/11/3/315.abstract.

CHAPTER 15: EATING DISORDERS

1. G. C. Patton et al., "Onset of Adolescent Eating Disorders: Population-Based Cohort Study Over 3 Years," *British Medical Journal* 318 (1999): 765–68.

2. M. L. Portella de Santana et al., "Epidemiology and Risk Factors of Eating Disorder in Adolescence: A Review," *Nutrición Hospitalaria* 27, no. 2 (2012): 391–401.

3. R. A. Gordon, *Eating Disorders: Anatomy of a Social Epidemic* (Oxford: Blackwell, 2000).

4. N. Wolf, *The Beauty Myth: How Images of Beauty Are Used Against Women* (New York: Dutton, 1991).

5. M. Nasser, "Comparative Studies of the Prevalence of Abnormal Eating Attitudes Among Arab Female Students at Both London and Cairo Universities," *Psychological Medicine* 16, no. 3 (1986): 621–25.

6. M. M. Fichter et al., "Time Trends in Eating Disturbances in Young Greek Migrants," *International Journal of Eating Disorders* 38, no. 4 (December 2005): 310–22.

7. S. J. Paxton et al., "Friendship Clique and Peer Influences on Body Image Concerns, Dietary Restraint, Extreme Weight Loss Behaviours and Binge Eating in Adolescent Girls," *Journal of Abnormal Psychology* 108 (1999): 255–66.

8. M. R. Hebl and T. F. Heatherton, "The Stigma of Obesity in Women: The Difference Is Black and White," *Personality and Social Psychology Bulletin* 24 (1998): 417–26.

9. General—Mervat Nasser, Melanie A. Katzman, and Richard A. Gordon, *Eating Disorders and Cultures in Transition* (New York: Routledge, 2001).

PART IV: THE DARK HEART OF PSEUDOSCIENCE
CHAPTER 16: RELATIVE RISK

1. V. Bouvard et al., "Carcinogenicity of Consumption of Red and Processed Meat," *Lancet* 16, no. 16 (December 2015): 1599–1600, thelancet.com/journals/lanonc/article/PIIS1470-2045%2815%2900444-1/fulltext.

CHAPTER 17: THE GAPS DIET

1. Katrin Starcke and Matthias Brand, "Decision Making Under Stress: A Selective Review," *Neuroscience and Biobehaviour Reviews* 36, no. 4 (2012): 1228–48.
2. F. Happé, A. Ronald, and R. Plomin, "Time to Give Up on a Single Explanation for Autism," *Nature Neuroscience* 9 (2006): 1218–20.
3. J. K. Hou, D. Lee and J. Lewis, "Diet and Inflammatory Bowel Disease: Review of Patient-Targeted Recommendations," *Clinical Gastroenterology and Hepatology* 12, no. 10 (2014): 1592–1600.
4. Zoe Connor, "What's Up with the GAPS Diet?" zoeconnor.co.uk/2016/05/whats-gaps-diet.

CHAPTER 18: CANCER

1. "Zen Macrobiotic Diets," *JAMA* 218, no. 3 (1971), jamanetwork.com/journals/jama/article-abstract/339481.
2. J. A. Chabot et al., "Pancreatic Proteolytic Enzyme Therapy Compared with Gemcitabine-Based Chemotherapy for the Treatment of Pancreatic Cancer," *Journal of Clinical Oncology* 28, no. 12 (2010): 2058–63.
3. American Cancer Society, "Questionable Methods of Cancer Management: 'Nutritional' Therapies," *CA: A Cancer Journal for Clinicians* 43 (1993): 309–19.
4. Rafael Moreno-Sánchez et al., "Energy Metabolism in Tumor Cells," *FEBS Journal* 274 (2007): 1393–1418.
5. Heather R. Christofk et al., "The M2 Splice Isoform of Pyruvate Kinase Is Important for Cancer Metabolism and Tumour Growth," *Nature* 452, no. 7184 (2008): 230–33.
6. Ken Garber, "Energy Deregulation: Licensing Tumors to Grow," *Science* 312, no. 5777 (2006): 1158–59.
7. General—Siddhartha Mukherjee, *The Emperor of All Maladies: A Biography of Cancer* (New York: Scribner's, 2010).
8. General—American Cancer Society, "Questionable Methods of Cancer Management—Nutritional Therapies," *CA: A Cancer Journal for Clinicians* 43 (1993): 300–19.
9. General—Dr. David Gorski's blogs from the always excellent Science-Based Medicine, sciencebasedmedicine.org/author/david-gorski.

PART V: THE FIGHT BACK
CHAPTER 19: THE EVOLUTION OF MYTHS

1. Chip Heath and Dan Heath, *Made to Stick: Why Some Ideas Survive and Others Die* (New York: Random House, 2007).

CHAPTER 21: FIGHTING PSEUDOSCIENCE

1. John Cook and Stephan Lewandowsky, *The Debunking Handbook* (2011), skepticalscience.com/Debunking-Handbook-now-freely-available-download.html.
2. Sydney E. Scott, Yoel Inbar, and Paul Rozin, "Evidence for Absolute Moral Opposition to Genetically Modified Food in the United States," *Perspectives on Psychological Science* 11, no. 3 (2016): 315–24, pps.sagepub.com/content/11/3/315.abstract.

About the Author

ANTHONY WARNER is a professional chef and blogger and a regular contributor to *New Scientist*, the *Sunday Times*, and The Pool. His blog has been featured in the *Guardian*, *Mail on Sunday*, *Telegraph*, and other publications. He lives in Nottinghamshire and blogs at angry-chef.com, and you can follow him @One_Angry_Chef.